Newnes
Engineering Materials
Pocket Book
Third Edition

Newnes

Engineering Materials Pocket Book

Third edition

W. Bolton

Newnes

OXFORD AUCKLAND BOSTON JOHANNESBURG MELBOURNE NEW DELHI

Newnes
An imprint of Butterworth-Heinemann
Linacre House, Jordan Hill, Oxford OX2 8DP
225 Wildwood Avenue, Woburn, MA 01801-2041

A member of the Reed Elsevier plc group

First published 1989
Second edition 1996
Third edition 2000

British Library Cataloguing in Publication Data
A catalogue record for this book is available from the British Library

ISBN 0 7506 4974 7

Library of Congress Cataloguing in Publication Data
A catalogue record for this book is available from the Library of Congress

Printed and bound in India at Indira Printers, New Delhi-110 020.

FOR EVERY TITLE THAT WE PUBLISH, BUTTERWORTH-HEINEMANN
WILL PAY FOR BTCV TO PLANT AND CARE FOR A TREE.

Contents

Preface to third edition

The main aim of this book is to provide engineers and students with a concise, pocket-size, affordable guide to the full range of materials used in engineering: ferrous and non-ferrous metals, polymeric materials, ceramics and composites. For the second edition the text has been extended to include chapters which group materials according to properties, this aiding the selection of materials required for having specific properties. The book is seen as being particularly useful to students engaged on project work. Obviously no book this size, or even any single book, can be completely exhaustive and so the selection of materials has been restricted to those most commonly encountered in engineering and those details of properties most relevant to general use of the materials. The book is not intended to replace the more detailed specifications given by the national and international standards groups.

To aid in interpreting the information given for the properties of materials, Chapter 1 gives a brief description of the main terms likely to be encountered, Chapter 2 gives an outline of the main testing methods and the Appendix conversion factors and tables for the various forms of units used to describe properties. Chapters 3 to 11 are devoted to data on the main engineering materials: ferrous, aluminium, copper, magnesium, nickel and titanium alloys, polymeric, ceramic and composite materials. In most cases, the chapters are broken down into five sections: a discussion of the materials, details of coding systems and compositions, heat treatment information, the properties of the materials, and typical uses to which they have been put. The codes and data given are for both American and British standards. To aid in selection, Chapters 12 to 16 group materials with properties: electrical, magnetic, mechanical, corrosion and wear resistance and thermal. Chapter 17 indicates the basis of selection methods. Chapter 19 is a consideration of the characteristics of the various types of processes which can be used to form products.

The main change from the second edition is the inclusion of a chapter on the selection of processes.

The book is essentially concerned with properties and there is only a very brief indication of the science of materials. For further information on materials science the reader is referred to textbooks, such as:

Anderson, J.C., Leaver, K.D., Rawlings, R.D., and Alexander, J.M., (1985). **Materials Science** (Van Nostrand, 3rd Ed.)
Ashby, M. and Jones, R.H. **Engineering Materials, vol. 1 and vol. 2** (Butterworth-Heinemann 1996, 1998)
Bolton, W, (1998). **Engineering Materials Technology** (Butterworth-Heinemann, 3rd Ed.)
Mills, N.N. (1986). **Plastics: Microstructure, Properties and Applications** (Arnold)
Smith, W.F. (1981). **Structure and Properties of Engineering Alloys** (McGraw- Hill)

For a discussion of selection procedures and case studies of materials selection, the reader is referred to:

Ashby, M.F. (1999). **Materials Selection in Mechanical Design** (Butterworth-Heinemann)
Bolton, W. (1998). **Engineering Materials Technology** (Butterworth-Heinemann, 3rd Ed.)
Crane, F.A.A. and Charles, J.A. (1997). **Selection and Use of Engineering Materials** (Butterworths-Heinemann 3rd Ed)
Farag, M.M. (1989). **Selection of Materials and Manufacturing Processes for Engineering Design** (Prentice Hall)

For a discussion of process selection the reader is referred to:

Swift K.G and Booker J.D. **Process Selection** (Arnold 1997)

The data used in this book has been obtained from a wide variety of sources. The main sources are:

The publications of the British Standards Institution
The publications of the American Society of Metals
Materials' manufacturers
Trade Associations

The interpretation and presentation of the data is however mine and should not be deemed to be that of any other organisation. For full details of standards, the reader is advised to consult the appropriate publications of the standards setting group.

W. Bolton

Acknowledgements

Extracts from British Standards are reproduced with the permission of BSI. Complete copies of the documents can be obtained by post from BSI Sales, Linford Wood, Milton Keynes, Bucks MK16 6LE.

Extracts from British Standards are reproduced with the permission of BSI. Complete copies of the Standards can be obtained by post from BSI Sales, Linford Wood, Milton Keynes MK14 6LE.

1 Terminology

The following, in alphabetical order, are definitions of the common terms used in engineering in connection with the properties of materials.

Additives. Plastics and rubbers almost invariably contain, in addition to the polymer or polymers, other materials, i.e additives. These are added to modify the properties and cost of the material.

Ageing. This term is used to describe a change in properties that occurs with certain metals at ambient or moderately elevated temperatures after hot working, a heat treatment process or cold working. The change is generally due to precipitation occurring, there being no change in chemical composition.

Alloy. This is a metallic material composed of two or more elements of which at least one is a metal.

Amorphous. An amorphous material is a non-crystalline material, i.e. it has a structure which is not orderly.

Annealing. This involves heating to, and holding at, a temperature which is high enough to result in a softened state for a material after a suitable rate of cooling, generally slowly. In the case of ferrous alloys the required temperature is the upper critical temperature. The purpose of annealing can be to facilitate cold working, improve machinability, improve mechanical properties, etc.

Anodizing. This term is used to describe the process, generally with aluminium, whereby a coating is produced on the surface of the metal by converting it to an oxide.

Atactic structure. A polymer structure in which side groups, such as CH_3, are arranged randomly on either side of the molecular chain.

Austempering. This is a heat treatment used with ferrous alloys. The material is heated to austenizing temperature and then quenched to the M_s temperature at such a rate that ferrite or pearlite is not formed. It is held at the M_s temperature until the transformation to bainite is complete.

Austenite. This term describes the structure of a solid solution of one or more elements in a face-centred cubic iron crystalline structure. It usually refers to the solid solution of carbon in the face-centred iron.

Austenitizing. This is when a ferrous alloy is heated to a temperature at which the transformation of its structure to austenite occurs.

Bainite. This describes a form of ferrite–cementite structure consisting of ferrite plates between which, or inside which, short cementite rods form, and occurs when ferrous alloys are cooled from the austenitic state at an appropriate rate of cooling. It is a harder structure than would be obtained by annealing but softer than martensite. The process used is called austempering.

Bend, angle of. The results of a bend test on a material are specified in terms of the angle through which the material can be bent without breaking (Figure 1.1). The greater the angle the more ductile the material. See Bend test, Chapter 2.

Figure 1.1 The angle of bend

Brinell number. The Brinell number is the number given to a material as a result of a Brinell test (see Hardness measurement, Chapter 2) and is a measure of the hardness of a material. The larger the number the harder the material.

Brittle failure. With brittle failure a crack is initiated and propagates prior to any significant plastic deformation. The fracture surface of a metal with a brittle fracture is bright and granular due to the reflection of light from individual crystal surfaces. With polymeric materials the fracture surface may be smooth and glassy or somewhat splintered and irregular.

Brittle material. A brittle material shows little plastic deformation before fracture. The material used for a china teacup is brittle. Thus because there is little plastic deformation before breaking, a broken teacup can be stuck back together again to give a cup the same size and shape as the original.

Carburizing. This is a form of case hardening which results in a hard surface layer being produced with ferrous alloys. The treatment involves heating the alloy to the austenitic state in a carbon-rich atmosphere so that carbon diffuses into the surface layers, then quenching to convert the surface layers to martensite.

Case hardening. This term is used to describe processes in which by changing the composition of surface layers of ferrous alloys a hardened surface layer can be produced. See Carburizing and Nitriding.

Cementite. This is a compound formed between iron and carbon, often referred to as iron carbide. It is a hard and brittle material.

Charpy test value. The Charpy test, see Impact tests Chapter 2, is used to determine the response of a material to a high rate of loading and involves a test piece being struck a sudden blow. The results are expressed in terms of the amount of energy absorbed by the test piece when it breaks. The higher the test value the more ductile the material.

Cis structure. A polymer structure in which a curved carbon backbone is produced by bulky side groups, e.g. CH_3, which are grouped all on the same side of the backbone.

Compressive strength. The compressive strength is the maximum compressive stress a material can withstand before fracture.

Copolymer. This is a polymeric material produced by combining two or more monomers in a single polymer chain.

Corrosion resistance. This is the ability of a material to resist deterioration by chemical or electrochemical reaction with its immediate environment. There are many forms of corrosion and so there is no unique way of specifying the corrosion resistance of a material.

Creep. Creep is the continuing deformation of a material with the passage of time when it subject to a constant stress. For a particular material the creep behaviour depends on both the temperature and the initial stress, the behaviour also depending on the material concerned. See Creep tests, Chapter 2.

Creep modulus. The initial results of a creep test are generally represented as a series of graphs of strain against time for different levels of stress. From these graphs values, for a particular time, of strains at different stresses can be obtained. The resulting stress–strain values can be used to give a stress–strain graph for a particular time, such a graph being referred to as an isochronous stress–strain graph. The creep modulus is the stress divided by the strain, for a particular time. The modulus is not the same as the tensile modulus. See Creep tests, Chapter 2.

Creep strength. The creep strength is the stress required to produce a given strain in a given time.

Crystalline. This term is used to describe a structure in which there is a regular, orderly, arrangement of atoms or molecules.

Damping capacity. The damping capacity is an indicator of the ability of a material to suppress vibrations.

Density. Density is mass per unit volume.

Dielectric constant. See permittivity.

Dielectric strength. The dielectric strength is a measure of the highest potential difference an insulating material can withstand without electric breakdown.

$$\text{Dielectric strength} = \frac{\text{breakdown voltage}}{\text{insulator thickness}}.$$

Ductile failure. With ductile failure there is a considerable amount of plastic deformation prior to failure. With metals the fracture shows a typical cone and cup formation and the fracture surfaces are rough and fibrous in appearance.

Ductile materials. Ductile materials show a considerable amount of plastic deformation before breaking.

Elastic limit. The elastic limit is the maximum force or stress at which, on its removal, the material returns to its original dimensions. For many materials the elastic limit and the limit of proportionality are the same, the limit of proportionality being the maximum force for which the extension is proportional to the force or the maximum stress for which the strain is proportional to the stress. See Tensile tests, Chapter 2.

Electrical conductivity. The electrical conductivity is a measure of the electrical conductance of a material, the bigger the conductance the greater the current for a particular potential difference. The electrical conductivity is defined by

$$\text{conductivity} = \frac{\text{length}}{\text{resistance} \times \text{cross-sectional area}},$$

$$\text{conductance} = \frac{1}{\text{resistance}}.$$

Conductance has the unit of ohm^{-1} or mho, conductivity has the unit Ω^{-1} m^{-1}. The IACS specification of conductivity is based on 100%, corresponding to the conductivity of annealed copper at 20°C; all other materials are then expressed as a percentage of this value.

Electrical resistivity. The electrical resistivity is a measure of the electrical resistance of a material, being defined by

$$\text{resistivity} = \frac{\text{resistance} \times \text{cross-sectional area}}{\text{length}}.$$

Resistivity has the unit Ω m.

Endurance. The endurance is the number of stress cycles to cause failure. See Fatigue tests, Chapter 2.

Endurance limit. The endurance limit is the value of the stress for which a test specimen has a fatigue life of N cycles. See Fatigue tests, Chapter 2.

Equilibrium diagram. This diagram is, for metals, constructed from a large number of experiments, in which cooling curves are determined for the whole range of alloys in a group, and provides a forecast of the states that will be present when an alloy of a specific composition is heated or cooled to a specific temperature.

Expansion, coefficient of linear. The coefficient of linear expansion is a measure of the amount by which a unit length of a material will expand when the temperature rises by one degree. It is defined by

$$\text{linear expansivity} = \frac{\text{change in length}}{\text{length} \times \text{temperature change}}.$$

It has the unit °C^{-1} or K^{-1}.

Expansivity, linear. This is an alternative name for the coefficient of linear expansion.

Fatigue life. The fatigue life is the number of stress cycles to cause failure. See Fatigue tests, Chapter 2.

Fatigue limit. The fatigue limit is the value of the stress below which the material will endure an infinite number of cycles. See Fatigue tests, Chapter 2.

Fatigue strength. The fatigue strength at N cycles is the value of the stress under which a test specimen has a life of N cycles. See Fatigue tests, Chapter 2.

Ferrite. This is a solid solution of one or more elements in body-centred cubic iron. It is usually used for carbon in body-centred cubic iron. Ferrite is comparatively soft and ductile.

Fracture toughness. The plane strain fracture toughness or opening-mode fracture toughness, K_{1c}, represents a practical lower limit of fracture toughness and is an indicator of whether a crack will grow or not.

Friction, coefficient of. The coefficient of friction is the maximum value of the frictional force divided by the normal force. In the situation where an object is to be started into motion, the maximum frictional force is the force needed to start the object sliding. Where an object is already in motion, the frictional force is that needed to keep it moving with a constant velocity. This is less than the frictional force needed to start sliding and so there are two coefficients of friction, a static coefficient and a dynamic coefficient, with the static coefficient larger than the dynamic coefficient.

Full hard. This term is used to describe the temper of alloys. It corresponds to the cold-worked condition beyond which the material can no longer be worked.

Glass transition temperature. The glass transition temperature is the temperature at which a polymer changes from a rigid to a flexible material. The tensile modulus shows an abrupt change from the high value typical of a glass-like material to the low value of a rubber-like material.

Half hard. This term is used to describe the temper of alloys. It corresponds to the cold-worked condition half-way between soft and full hard.

Hardenability. The term hardenability of a material is used as a measure of the depth of hardening introduced into a material by quenching (see Hardenability, Chapter 2).

Hardening. This describes a heat treatment by which hardness is increased.

Hardness. The hardness of a material may be specified in terms of some standard test involving indentation, e.g. the Brinell, Vickers and Rockwell tests, or scratching of the surface of the material, the Moh test. See Hardness measurement, Chapter 2.

Heat distortion/deflection temperature. This is the temperature at which a strip of polymeric material under a specified load shows a specified amount of deflection.

Heat-resisting alloy. This is an alloy developed for use at high temperatures.

Homopolymer. This describes a polymer that has molecules made up of just one monomer.

Hooke's law. When a material obeys Hooke's law its extension is directly proportional to the applied stretching forces. See Tensile tests in Chapter 2.

Impact properties. See Charpy test value and Izod test value, also Impact tests in Chapter 2.

Isochronous stress–strain graph. See the entry on Creep modulus.

Isotactic structure. A polymer structure in which side groups of molecules are arranged all on the same side of the molecular chain.

Izod test value. The Izod test, see Impact tests, Chapter 2, is used to determine the response of a material to a high rate of loading and involves a test piece being struck a sudden blow. The results are expressed in terms of the amount of energy absorbed by the test piece when it breaks. The higher the test value the more ductile the material.

Jominy test. This is a test used to obtain information on the hardenability of alloys. See Chapter 2 for more information.

Limit of proportionality. Up to the limit of proportionality, the extension is directly proportional to the applied stretching forces, i.e. the strain is proportional to the applied stress (see Figure 1.4).

Machinability. There is no accepted standard test for machinability and so it is based on empirical test data and is hence subjective. Machinability is a measure of the differences encountered in machining a material.

Maraging. This is a precipitation hardening treatment used with some ferrous alloys. See Precipitation hardening.

Martensite. This is a general term used to describe a form of structure. In the case of ferrous alloys it is a structure produced when the rate of cooling from the austenitic state is too rapid to allow carbon atoms to diffuse out of the face-centred cubic form of austenite and produce the body-centred form of ferrite. The result is a highly strained hard structure.

Melting point. This is the temperature at which a material changes from solid to liquid.

Mer. See Monomer.

Moh scale. This is a scale of hardness arrived at by considering the ease of scratching a material. It is a scale of 10, with the higher the number the harder the material. See Impact tests, Chapter 2.

Monomer. This is the unit, or mer, consisting of a relatively few atoms which are joined together in large numbers to form a polymer.

Nitriding. This is a treatment in which nitrogen diffuses into surface layers of a ferrous alloy and hard nitrides are produced, hence a hard surface layer.

Normalizing. This heat treatment process involves heating a ferrous alloy to a temperature which produces a fully austenitic structure, followed by air cooling. The result is a softer material, but not as soft as would be produced by annealing.

Orientation. A polymeric material is said to have an orientation, uniaxial or biaxial, if during the processing of the material the molecules become aligned in particular directions. The properties of the material in such directions is markedly different from those in other directions.

Pearlite. This is a lamellar structure of ferrite and cementite.

Percentage elongation. The percentage elongation is a measure of the ductility of a material, the higher the percentage the greater the ductility. See Tensile tests, Chapter 2.

$$\text{Percentage elongation} = \frac{\text{final} - \text{initial lengths}}{\text{initial length}} \times 100$$

Percentage reduction in area. The percentage reduction in area is a measure of the ductility of a material, the higher the percentage the greater the ductility. See Tensile tests, Chapter 2.

$$\text{Percent. reduction in area} = \frac{\text{final} - \text{initial areas}}{\text{initial area}} \times 100$$

Permeability. This term is used to describe the rate at which gases or vapours are transmitted through a material. The rate of transmis sion per unit surface area of the material is given by:

$$\text{rate of transmission/area} = P\,(p_1 - p_2)/L$$

where P is the permeability coefficient, p_1 and p_2 the pressures on each side of the material and L the material thickness. A variety of units are used for permeability values. In some the pressure is quoted in centimetres of mercury, in others in Pa or N m^{-2}. The time might be in days or seconds. The rate may be quoted in terms of volumes in cubic centimetres (cm^3) or mass in moles or kilograms (kg) or grams (g).

Permittivity. The relative permittivity ϵ_r, or dielectric constant, of a material can be defined as the ratio of the capacitance of a capacitor with the material between its plates compared with that of the same capacitor with a vacuum.

$$\epsilon = \epsilon_r\,\epsilon_0$$

where ϵ is the absolute permittivity and ϵ_0 the permittivity of free space, i.e. a vacuum.

Plane strain fracture toughness. See fracture toughness.

Poisson's ratio. Poisson's ratio is the ratio (transverse strain)/(longitudinal strain).

Precipitation hardening. This is a heat treatment process which results in a precipitate being produced in such a way that a harder material is produced.

Proof stress. The 0.2% proof stress is defined as that stress which results in a 0.2% offset, i.e. the stress given by a line drawn on the stress–strain graph parallel to the linear part of the graph and passing through the 0.2% strain value (Figure 1.2). The 0.1% proof stress is similarly defined. Proof stresses are quoted when a material has no well defined yield point. See Tensile tests, Chapter 2.

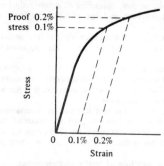

Figure 1.2 Determination of proof stress

Quenching. This is the method used to produce rapid cooling.

Recovery, fractional. The fractional recovery is defined as the strain recovered divided by the creep strain, when the load is removed.

Recrystallization. This is generally used to describe the process whereby a new, strain free grain structure is produced from that existing in a cold-worked metal by heating.

Refractive index. The refractive index of a material is the ratio (speed of light in a vacuum)/(speed of light in the material). For some materials the speed of light depends on the direction through the material the light is traversing and so the refractive index varies with direction.

Relative permeability. This is a measure of the magnetic properties of a material, being defined as the ratio of the magnetic flux density in the material to the flux density in a similar situation when the material is replaced by a vacuum.

Relative permittivity. See permittivity.

Resilience. This term is used with elastomers to give a measure of the elasticity of a material. A high resilience material will suffer elastic collisions, when a high percentage of the kinetic energy before the collision is returned to the object after the collision. A less resilient material would loose more kinetic energy in the collision.

Rigidity, modulus of. The modulus of rigidity is the slope of the shear stress/shear strain graph below the limit of proportionality.

Rockwell test value. The Rockwell test is used to give a value for the hardness of a material. There are a number of Rockwell scales and thus the scale being used must be quoted with all test results (see Impact tests, Chapter 2).

Ruling section. The limiting ruling section is the maximum diameter of round bar, at the centre of which the specified properties may be obtained.

Rupture stress. The rupture stress is the stress to cause rupture in a given time at a given temperature and is widely used to describe the creep properties of materials. See Creep tests, Chapter 2.

Secant modulus. For many polymeric materials there is no linear part of the stress–strain graph and thus a tensile modulus cannot be quoted. In such cases the secant modulus is used. It is the stress at a value of 0.2% strain divided by that strain (Figure 1.3).

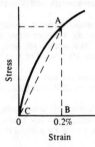

Figure 1.3 The secant modulus is AB/BC

Shear. When a material is loaded in such a way that one layer of the material is made to slide over an adjacent layer then the material is said to be in shear.

Shear strength. The shear strength is the shear stress required to produce fracture.

Shore durometer. This is a method for measuring the hardness of polymers and rubbers. A number of scales are used. See Chapter 2.

Sintering. This is the process by which powders are bonded by molecular or atomic attraction, as a result of heating to a temperature below the melting points of the constituent powders.

S/N graph. This is a graph of the stress amplitude S plotted against the number of cycles N for the results from a fatigue test. The stress amplitude is half the algebraic difference between the maximum and minimum stresses to which the material is subject. See Fatigue tests, Chapter 2.

Solution treatment. This heat treatment involves heating an alloy to a suitable temperature, holding at that temperature long enough for one or more constituent elements to enter into solid solution, and then cooling rapidly enough for these to remain in solid solution.

Specific gravity. The specific gravity of a material is the ratio of its density compared with that of water.

$$\text{Specific gravity} = \frac{\text{density of material}}{\text{density of water}}$$

Specific heat capacity. The amount by which the temperature rises for a material, when there is a heat input depends on its specific heat capacity. The higher the specific heat capacity the smaller the rise in temperature per unit mass for a given heat input.

$$\text{Specific heat capacity} = \frac{\text{heat input}}{\text{mass} \times \text{change in temperature}}.$$

Specific heat capacity has the unit J kg^{-1} K^{-1}.

Spheroidizing. This is a treatment used to produce spherical or globular forms of carbide in steel.

Strain. The engineering strain is defined as the ratio (change in length)/(original length) when a material is subject to tensile or compressive forces. Shear strain is the ratio (amount by which one layer slides over another)/(separation of the layers). Because it is a ratio, strain has no units, though it is often expressed as a percentage. Shear strain is usually quoted as an angle in radians.

Strain hardening. This is an increase in hardness and strength produced as a result of plastic deformation at temperatures below the recrystallization temperature, i.e. cold working.

Strength. See Compressive strength, Shear strength and Tensile strength.

Stress. In engineering tensile and compressive stress is usually defined as (force)/(initial cross-sectional area). The true stress is (force)/(cross-sectional area at that force). Shear stress is the (shear force)/(area resisting shear). Stress has the unit Pa (pascal) or N m^{-2} with 1 Pa = 1 N m^{-2}.

Stress relieving. This is a treatment to reduce residual stresses by heating the material to a suitable temperature, followed by slow cooling.

Stress–strain graph. The stress–strain graph is usually drawn using the engineering stress (see Stress) and engineering strain (see Strain). Figure 1.4 shows an example of the form one takes for a metal like mild steel. See Tensile tests, Chapter 2.

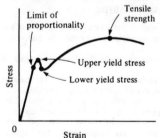

Figure 1.4 Stress–strain graph

Surface hardening. This is a general term used to describe a range of processes by which the surface of a ferrous alloy is made harder than its core.

Syndiatic structure. A polymer structure in which side molecular groups are arranged in a regular manner, alternating from one side to the other of the molecular chain.

Temper. This term is used with non-ferrous alloys as an indication of the degree of hardness/strength, with expressions such as hard, half-hard, three-quarters hard being used.

Tempering. This is the heating of a previously quenched material to produce an increase in ductility.

Tensile modulus. The tensile modulus, or Young's modulus, is the slope of the stress–strain graph over its initial straight-line region (Figure 1.5). See Tensile tests, Chapter 2.

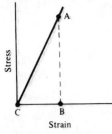

Figure 1.5 The tensile modulus is AB/BC

Tensile strength. This is defined as (maximum force before breaking)/ (initial cross-sectional area). See Figure 1.4. See Tensile tests, Chapter 2.

Thermal conductivity. The rate at which energy is transmitted as heat through a material depends on a property called the thermal conductivity. The higher the thermal conductivity the greater the rate at which heat is conducted. Thermal conductivity is defined by

$$\text{thermal conductivity} = \frac{\text{rate of transfer of heat}}{\text{cross-sectional area} \times \text{temp. gradient}}.$$

Thermal conductivity has the unit $W\ m^{-2}\ K^{-1}$.

Thermal expansivity. See Expansion, coefficient of linear.

Transmission factor, direct. The direct transmission factor is the ratio (transmitted light flux)/(incident light flux). It is usually expressed as a percentage. Because the transmission factor depends on the thickness of the material, results are usually standardized to a thickness of 1 mm. For some materials this can mean a transmission factor of virtually 0 % for 1 mm when they are not completely opaque for smaller thicknesses.

Transition temperature. The transition temperature is the temperature at which a material changes from giving a ductile failure to giving a brittle failure.

Trans structure. A polymer structure in which a relatively straight stiff carbon backbone is produced by CH_3 groups alternating from one side to the other of the backbone.

Vickers' test results. The Vickers test is used to give a measure of hardness (see Hardness measurement, Chapter 2). The higher the Vickers hardness number the greater the hardness.

Water absorption. This is the percentage gain in weight of a polymeric material, after immersion in water for a specified amount of time, under controlled conditions.

Wear resistance. This is a subjective comparison of the wear resistance of materials. There is no standard test.

Weldability. The weldability of a particular combination of metals indicates the ease with which sound welds can be made.

Work hardening. This is the hardening of a material produced as a consequence of working, subjecting it to plastic deformation at temperatures below those of recrystallization.

Yield point. For many metals, when the stretching forces applied to a test piece are steadily increased, a point is reached when the extension is no longer proportional to the applied forces and the extension increases more rapidly than the force, until a maximum force is reached. This is called the upper yield point. The force then drops to a value called the lower yield point before increasing again as the extension is continued (see Figure 1.4). See Tensile tests, Chapter 2.

Young's modulus. See Tensile modulus.

2 Test methods

The following are some of the more common test methods used to determine the properties of materials used in engineering.

Bend tests

The bend test is a simple test of ductility. It involves bending a sample of the material through some angle and determining whether the material is unbroken and free from cracks after such a bend. The results of such a test are specified in terms of the angle of bend (see Figure 1.1).

Creep tests

For metals, other than very soft metals like lead, creep effects are negligible at ordinary temperatures but become significant at higher temperatures. For this reason, creep tests of metals are carried out generally at high temperatures. Figure 2.1 shows the essential features of such a test. The temperature of the test piece is kept constant during a test, being monitored by a thermocouple attached to the test piece. The extension is measured with time, tests often being carried out over quite protracted periods of time. For polymeric materials, creep is often quite significant at ordinary temperatures.

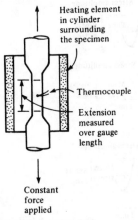

Figure 2.1 A creep test

The results of creep tests can be presented as a graph of strain plotted against time, a specification of the stress to rupture in a particular time (see Rupture stress, Chapter 1), or, in particular for polymeric materials, an isochronous stress–strain graph (see Creep modulus, Chapter 1) or a specification of the creep modulus (see Chapter 1).

Fatigue tests

Various fatigue tests have been devised to simulate the changes of stress to which the materials of different components are subjected when in service. Bending-stress machines are used to bend a test piece alternately one way and then the other (Figure 2.2a), whereas torsional-fatigue machines twist it alternately one way and then the opposite way (Figure 2.2b). Another type of machine is used to produce alternating tension and compression by direct stressing (Figure 2.2c). The tests can be carried out with stresses which alternate about zero stress (Figure 2.2d), apply a repeated stress which varies from zero to some maximum value (Figure 2.2e) or apply a stress which varies about some stress value and does not reach zero at all (Figure 2.2f). The aim of the test is to subject the material to the types of stresses to which it will be subjected in service.

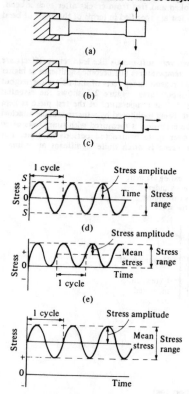

Figure 2.2 Fatigue testing (a) Bending, (b) torsion, (c) direct stress, (d) alternating stress, (e) repeated stress, (f) fluctuating stress

Fatigue test results can be expressed as an S/N graph (see Chapter 1), a specification of the fatigue limit (see Chapter 1) or the endurance limit for N cycles (see Chapter 1).

Hardenability

Hardenability is measured by the response of a standard test piece to a standard test, called the Jominy test. This involves heating the steel test piece to its austenitic state, fixing it in a vertical position and then quenching the lower end by means of a jet of water (Figure 2.3). This method of quenching results in different rates of cooling along the length of the test piece. When the test piece is cool, after the quenching, a flat portion is ground along one side, about 0.4 mm deep, and hardness measurements made along the length of the test piece. The significant point about the Jominy tests is not that they give the hardness at different positions along the test piece but that they give the hardness at different cooling rates, since each distance along the test piece corresponds to a different cooling rate (Figure 2.4). If the cooling rates are known at points, both on the surface and within a sample of steel, then the Jominy results can be used to indicate the hardness that will occur at those points.

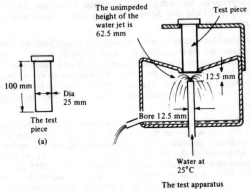

Figure 2.3 The Jominy test

Cooling rate at 700°C (°C s⁻¹)

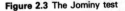

Distance from quenched end (mm)

Figure 2.4 Cooling rates at different distances from the quenched end of the Jominy test piece

Hardness measurement

The most common form of hardness measurements for metals involves standard indenters being pressed into the surface of the material concerned. Measurements associated with the indentation are then taken as a measurement of the hardness of the surface. The Brinell test, the Vickers test and the Rockwell test are the main forms of such tests. See Brinell test values, Vickers test values and Rockwell test values in Chapter 1.

With the *Brinell test*, a hardened steel ball is pressed for a time of 10 to 15 s into the surface of the material by a standard force. After the load and ball have been removed, the diameter of the indentation is measured. The Brinell hardness number, usually signified by

HB, is obtained by dividing the size of the force applied, by the spherical area of the indentation. This area can be obtained by calculation or the use of tables which relate the area to the diameter of the ball and the diameter of the indentation.

$$\text{Hardness} = \frac{\text{applied force}}{\text{spherical surface area of indent}}$$

The units used for the area are mm^2 and for the force kgf (1 kgf = 9.8 N). The diameter D of the ball used and the size of the applied force F are chosen to give F/D^2 values of 1, 5, 10 or 30, the diameters of the balls being 1, 2, 5 or 10 mm. In principle, the same value of F/D^2 will give the same hardness value, regardless of the diameter of the ball used.

The Brinell test cannot be used with very soft or very hard materials. In the one case the indentation becomes equal to the diameter of the ball, and in the other case there is either no or little indentation on which measurements can be made. The thickness of the material being tested should be at least ten times the depth of the indentation, if the results are not to be affected by the thickness of the sample.

The *Vickers* test uses a diamond indenter which is pressed for 10 to 15 s into the surface of the material under test. The result is a square-shaped impression. After the load and indenter are removed, the diagonals of the indentation are measured. The Vickers' hardness number, signified by HV, is obtained by dividing the size of the force applied by the surface area of the indentation. The surface area can be calculated, the indentation being assumed to be a right pyramid with a square base and a vertex angle of 136°, this being the vertex angle of the diamond. Alternatively, tables can be used to relate the diagonal values with the area.

The Vickers test has the advantage over the Brinell test of the increased accuracy that is possible in determining the diagonals of a square, as opposed to the diameter of a circle. Otherwise it has the same limitations as the Brinell test.

The *Rockwell test* uses either a diamond cone or a hardened steel ball as the indenter. A force of 90.8 N is applied to press the indenter into contact with the surface. A further force is then applied and causes an increase in depth of indenter penetration into the material. The additional force is then removed, and there is some reduction in the depth of the indenter due to the deformation of the material not being entirely plastic. The difference in the final depth of the indenter and the depth before the additional force was applied, is determined. This is the permanent increase in penetration (e) due to the additional force.

$$\text{Hardness} = E - e$$

where E is a constant determined by the form of the indenter. For the diamond cone indenter, E is 100; for the steel ball, E is 130.

There are a number of Rockwell scales, the scale being determined by the indenter and the additional force used. Table 2.1 indicates the scales and the types of materials for which each is typically used. In any reference to the results of a Rockwell test, signified by HR, the scale letter must be quoted. The B and C scales are probably the most commonly used for metals.

Table 2.1 Rockwell scales and typical applications

Scale	Indenter	Force (kN)	Typical applications
A	Diamond	0.59	Thin steel and shallow case-hardened steel
B	Ball 1.588 mm dia.	0.98	Copper alloys, aluminium alloys, soft steels
C	Diamond	1.47	Steel, hard cast irons, deep case-hardened steel
D	Diamond	0.98	Thin steel and medium case-hardened steel
E	Ball 3.175 mm dia.	0.98	Cast iron, aluminium, magnesium and bearing alloys
F	Ball 1.588 mm dia.	0.59	Annealed copper alloys, thin soft sheet metals, brass
G	Ball 1.588 mm dia.	1.47	Malleable irons, gun metals, bronzes, copper–nickel alloys
H	Ball 3.175 mm dia.	0.59	Aluminium, lead, zinc
K	Ball 3.175 mm dia.	1.47	Aluminium and magnesium alloys
L	Ball 6.350 mm dia.	0.59	Plastics
M	Ball 6.350 mm dia.	0.98	Plastics
P	Ball 6.350 mm dia.	1.47	
R	Ball 12.70 mm dia.	0.59	Plastics
S	Ball 12.70 mm dia.	0.98	
V	Ball 12.70 mm dia.	1.47	

For most commonly used indenters with the Rockwell test, the size of the indentation is rather small. Thus localized variations of structure and composition can affect the result. However, unlike the Brinell and Vickers tests flat, polished surfaces are not required for accurate measurements.

The standard Rockwell test cannot be used with thin sheet; however a variation of the test, known as the Rockwell superficial hardness test, can be used. Smaller forces are used and the depth of indentation is determined with a more sensitive device, as much smaller indentations are used. An initial force of 29.4 N is used instead of 90.8 N. Table 2.2 lists the scales given by this test.

Table 2.2 Rockwell scales for superficial hardness

Scale	Indenter	Additional force (kN)
15-N	Diamond	0.14
30-N	Diamond	0.29
45-N	Diamond	0.44
15-T	Ball 1.588 mm dia.	0.14
30-T	Ball 1.588 mm dia.	0.29
45-T	Ball 1.588 mm dia.	0.44

The Brinell, Vickers and Rockwell tests can be used with polymeric materials. The Rockwell test, with its measurement of penetration, rather than surface area of indentation, is more widely used. Scale R is a commonly used scale.

The *Shore durometer* is used for measuring the hardness of polymers and elastomers, giving hardness values on a number of Shore scales. For the A scale, a truncated cone indenter is pressed against the material surface by a load of 8 N. The depth of the penetration of the indenter is measured. This must be done immediately the load is applied since the value will change with time. For scale D, a rounded tip cone is used with a load of 44.5 N.

One form of hardness test is based on assessing the resistance of a material to being scratched. The *Moh* scale consists of ten materials arranged so that each one will scratch the one preceding it in the scale, but not the one that succeeds it.

1 Talc
2 Gypsum
3 Calcspar
4 Fluorspar
5 Apatite
6 Felspar
7 Quartz
8 Topaz
9 Corundum
10 Diamond

Ten styli of the materials in the scale are used for the test. The hardness number of a material under test is one number less than that of the substance that just scratches it.

Figure 2.5 shows the general range of hardness values for different materials when related to the Vickers, Brinell, Rockwell and Moh hardness scales. There is an approximate relationship between hardness values and tensile strengths. Thus for annealed steels the tensile strength in MPa (MN m^{-2}) is about 3.54 times the Brinell hardness value, and for quenched and tempered steel 3.24 times the Brinell hardness value. For brass the factor is about 5.6, and for aluminium alloys about 4.2.

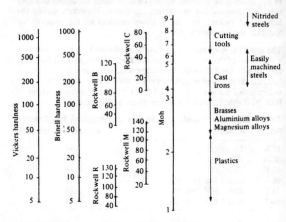

Figure 2.5 Hardness scales and typical values

Impact tests

There are two main forms of impact test, the Izod and Charpy tests (see Charpy test value and Izod test value, Chapter 1). Both tests involve the same type of measurement but differ in the form of the test piece. Both involve a pendulum swinging down from a specified height to hit the test piece (Figure 2.6). The height to which the pendulum rises after striking and breaking the test piece, is a measure of the energy used in the breaking. If no energy were used, the pendulum would swing up to the same height as it started from. The greater the energy used in the breaking, the lower the height to which the pendulum rises. Both the American and British specifications require the same standard size test pieces.

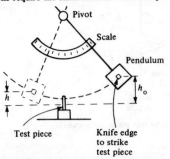

Figure 2.6 The principle of impact testing

With the *Izod* test the energy absorbed in breaking a cantilevered test piece (Figure 2.7) is measured. The test piece is notched on one face and the blow is struck on the same face, at a fixed height above the notch. The test pieces are, for metals, either 10 mm square or 11.4 mm diameter if they conform to British Standards. Figure 2.8 shows the details of the 10 mm square test piece. For polymeric materials, the standard test pieces are either 12.7 mm square (Figure 2.9) or 12.7 mm by 6.4 to 12.7 mm, depending on the thickness of the material concerned. With metals the pendulum strikes the test piece at a speed between 3 and 4 m s^{-1}, with polymeric materials the speed is 2.44 m s^{-1}.

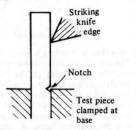

Figure 2.7 Form of the Izod test piece

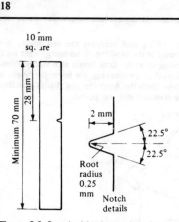

Figure 2.8 Standard Izod test piece for a metal

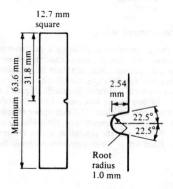

Figure 2.9 Standard Izod test piece for a polymeric material

With the *Charpy* test the energy absorbed in breaking a beam test piece (Figure 2.10) is measured. The test piece is supported at each end and is notched in the middle, the notch being on the face directly opposite to where the pendulum strikes the test piece. For metals, the British Standard test piece has a square cross-section of side 10 mm and a length of 55 mm. Figure 2.11 shows the details of the standard test piece and the three forms of notch that are possible. The results obtained with the different forms of notch cannot be compared, thus for the purpose of comparison between metals the same type of notch should be used. The test pieces for polymeric materials are tested either in the notched or unnotched state. A standard test piece is 120 mm long, 15 mm wide and 10 mm thick in the case of moulded polymeric materials. Different widths and thicknesses are used with sheet polymeric materials. The notch is produced by milling a slot across one face. The slot has a width of 2 mm and a radius of less than 0.2 mm at the corners of the base and walls of the slot. With metals, the pendulum strikes the test piece at a speed between 3 and 5.5 m s^{-1}, with polymeric materials, the speed is between 2.9 and 3.8 m s^{-1}.

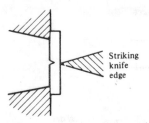

Figure 2.10 Form of the Charpy test piece

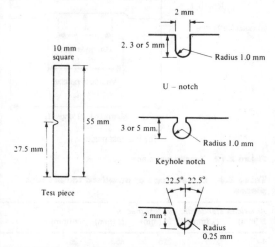

Figure 2.11 Standard Charpy test piece for a metal

The results of impact tests need to specify not only the type of test, i.e. Izod or Charpy, but the form of notch used. In the case of metals, the results are expressed as the amount of energy absorbed by the test piece when it breaks. In the case of polymeric materials, the results are often given as absorbed energy divided by either the cross-sectional area of the unnotched test piece, or the cross-sectional area behind the notch in the case of notched test pieces.

Tensile test

In the tensile test, measurements are made of the force required to extend a standard size test piece at a constant rate, the elongation of a specified gauge length of the test piece being measured by some form of extensometer. In order to eliminate any variations in data obtained from the test, due to differences in shapes of test pieces, standard shape and size test pieces are used. Figure 2.12 shows the forms of two standard test pieces, one being a flat test piece and the other a round test piece. The dimensions of such standard test pieces are given in Table 2.3. These apply to all metals other than cast irons. An important feature of the dimensions is the radius given for the shoulders of the test pieces. Variations in the radii can markedly affect the data obtained from a test.

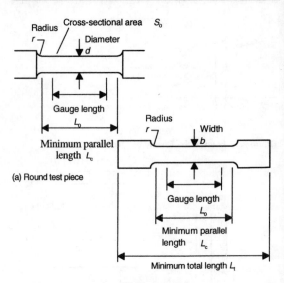

(a) Round test piece

(b) Flat test piece

Figure 2.12 Tensile test pieces

Table 2.3 Dimensions of standard tensile test pieces

American standards: flat test pieces

b (mm)	L_0 (mm)	L_c (mm)	L_t (mm)	r (mm)
40	200	225	450	25
12.5	50	60	200	13
6.25	25	32	100	6

American standards: round test pieces

d (mm)	L_0 (mm)	L_c (mm)	r (mm)
12.50	50	60	10
8.75	35	45	6
6.25	25	32	5
4.00	16	20	4
2.50	10	16	2

Note: the gauge length $L_0 = = 04d$.

European standards

For proportional test pieces $L_0 = k\sqrt{S_0}$, where S_0 is the cross-sectional area. The cross-section may be circular, square, rectangular, annular or in special cases some other shape. k is 5.65. When the cross-sectional area is to small for this requirement to be met, a higher value of $k = 11.3$ may be used. For non-proportional test pieces from sheet, strip or flats 0.1 to 0.3 mm thick:

Test piece type	Width b (mm)	L_0 (mm)	L_c (mm)	L_t (mm)ı
1	12.5 ± 150	75	87.5	
2	20 ± 1	80	120	140

Note: * refers to parallel sides test pieces.

For sheets and flats of thickness equal to or greater than 3 mm:

For circular cross-section: $L_c \geq L_0 = d/2$

For prismatic cross-sections: $L_c \geq L_0 = 1.5\sqrt{S_0}$

For proportional test pieces $L_0 = 5.65\sqrt{S_0}$. This gives, for circular cross-sections, $L_0 = 5d$. For such test pieces:

d (mm)	S_0 (mm²)	L_0 (mm)	L_c (mm)
20 ± 0.150	314.2	100 ± 1.0	110
10 ± 0.075	78.5	50 ± 0.5	55
5 ± 0.040	19.6	25 ± 0.25	28

Note: $L_t > L_c = 2d$.

The rate of stressing for the European test pieces should be:

Tensile modulus (GPa)	Rate (MPa/s)	
	Minimum	Maximum
<150	2	10
≥150	6	30

3 Ferrous alloys

3.1 Materials

Alloys

The term ferrous alloys is used for all those alloys having iron as the major constituent. Pure iron is a relatively soft material and is hardly of any commercial use in that state. Alloys of iron with carbon are classified according to their carbon content as shown in Table 3.1.

Table 3.1 Alloys of iron with carbon

Material	Percentage carbon
Wrought iron	0 to 0.05
Steel	0.05 to 2
Cast iron	2 to 4.3

The term carbon steel is used for those steels in which essentially just iron and carbon are present. The term alloy steel is used where other elements are included. Stainless steels are one form of alloy steel which has high percentages of chromium in order to give it a high resistance to corrosion. The term tool steels is used to describe those steels, carbon or alloy, which are capable of being hardened and tempered and have suitable properties for use as a tool material.

The following is an alphabetical listing of the various types of ferrous alloys.

Alloy steels

The term low alloy is used for alloy steels when the alloying additions are less than 2%, medium alloy between 2 and 10% and high alloy when over 10%. In all cases the amount of carbon is less than 1%. Common elements that are added are aluminium, chromium, cobalt, copper, lead, manganese, molybdenum, nickel, phosphorus, silicon, sulphur, titanium, tungsten and vanadium.

There are a number of ways in which the alloying elements can have an effect on the properties of the steel. The main effects are to:

1 Solution harden the steel
2 Form carbides
3 Form graphite
4 Stabilize austenite or ferrite
5 Change the critical cooling rate
6 Improve corrosion resistance
7 Change grain growth
8 Improve machinability

See Coding system for steels, Composition of alloy steels, Creep properties, Machinability, Oxidation resistance , Mechanical properties of alloy steels, Thermal properties and Uses of alloy steels.

Carbon steels

The term carbon steel is used for those steels in which essentially just iron and carbon are present. Such steels with less than 0.80% carbon are called hypo-eutectoid steels, those with between 0.80% and 2.0% carbon being hyper-eutectoid steels. Steels with between 0.10% and 0.25% carbon are termed mild steels, between 0.20% and 0.50% medium-carbon steels and more than 0.50% high carbon steels.

Figure 3.1 shows the iron–carbon equilibrium diagram.

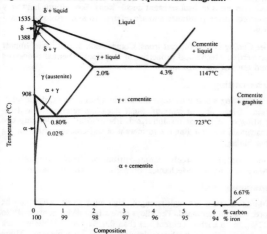

Figure 3.1 The iron-carbon equilibrium diagram

See Coding system for steels, Composition of carbon steels, Creep properties, Hardness, Impact properties, Machinability, Mechanical properties of carbon steels, Thermal properties and Uses of carbon steels.

Cast irons

Cast irons can be divided into five main categories:

1 Grey irons
These contain graphite in the form of flakes. The most widely used form has the graphite in a pearlitic structure.

2 Ductile irons or spheroidal-graphite irons
These contain graphite in the form of nodules as a result of magnesium or cerium being added during casting. The material is more ductile than grey irons.

3 White irons
There is no graphite present, the iron containing hard cementite. This material is hard and brittle, hence difficult to machine. The main use is where a wear-resistant surface is required.

4 Malleable irons
These are produced by the heat treatment of white irons. These are sometimes referred to as falling into two categories, ferritic or pearlitic, or considered as three groups, whiteheart, blackheart or pearlitic. Malleable irons have better ductility than grey cast irons and this, combined with their high tensile strength, makes them a widely used material.

5 High alloy irons
These are alloys containing significant percentages of elements such as silicon, chromium, nickel or aluminium. They can be considered to fall into two categories, graphite-free white irons or graphite-containing irons. The graphite-free white irons are very abrasion resistant. In the graphite-containing irons, the graphite is in the form

of flakes or nodules, hence the terms sometimes used of heat resistant grey irons or heat-resistant ductile irons. Some forms have very good corrosion resistance and are referred to as corrosion-resistant irons.

See Coding system for cast irons, Composition of cast irons, Impact properties, Mechanical properties of cast irons, Thermal properties and Uses of cast irons.

Free-cutting steels

Free-cutting steels have high machinability as a result of the addition of lead and/or sulphur. Such steels are referred to as free-cutting or leaded or resulphurized. Phosphorus can also improve machinability by aiding the formation of self-breaking chips during machining.

See Coding of steels, Composition of free-cutting steels, Machinability and Mechanical properties of free-cutting steels.

Maraging steels

Maraging steels are high strength, high alloy steels, which can be precipitation hardened. The alloys have a high nickel content, 18 to 22%, and a carbon content less than 0.03%. Other elements such as cobalt, titanium and molybdenum are also present. These elements form intermetallic compounds with the nickel. The carbon content is kept low, since otherwise the high nickel content could lead to the formation of graphite in the structure and a consequential drop in strength and hardness. Typically the heat treatment consists of heating to about 830°C and then air cooling. This results in a martensitic structure. Following machining and working, the steel is then precipitation hardened by heating to about 500°C for two or three hours. Typically, prior to the precipitation treatment the material might have a tensile strength of about 700 MPa or MN m^{-2} and a hardness of 300 HV, while afterwards it is about 1700 MPa or MN m^{-2} and 550 HV.

See Composition of maraging steels and Mechanical properties of maraging steels.

Stainless steels

There are several types of stainless steel: ferritic, martensitic and austenitic. Ferritic steels contain between 12 and 25% chromium and less than 0.1% carbon. Such steels, on cooling from the liquid, only change to ferrite and thus, since no austenite is formed, hardening by quenching to give martensite cannot occur. Such steels can however be hardened by cold working. Martensitic steels contain between about 12 and 18% chromium and 0.1 to 1.2% carbon. Such a steel, when cooled from the liquid state, produces austenite and so can be hardened by quenching to give martensite. Martensitic steels are subdivided into three groups: stainless irons, stainless steels and high chromium steels. Stainless irons contain about 0.1% carbon and 12 to 13% chromium, stainless steels about 0.25 to 0.30% carbon with 11 to 13% chromium, and high chromium steels about 0.05 to 0.15% carbon with 16 to 18% chromium and 2% nickel. Austenitic steels contain 16 to 26% chromium, more than about 6% nickel and very low percentages of carbon, 0.1% or less. Such alloys are completely austenitic at all temperatures. They cannot be hardened by quenching, but can be by cold working.

During welding, stainless steels may undergo structural changes which are detrimental to the corrosion resistance of the material.

The effect is known as weld decay and results from the precipitation of chromium rich carbides at grain boundaries. One way to overcome this is to stabilize the steel by adding other elements, such as niobium and titanium, which have a greater affinity for the carbon than the chromium and so form carbides in preference to the chromium.

See Coding system for stainless steels, Composition of stainless steels, Creep properties, Oxidation resistance, Mechanical properties of stainless steels, Thermal properties and Uses of stainless steels.

Tool steels

Plain carbon steels obtain their hardness from their high carbon content, the steels needing to be quenched in cold water to obtain maximum hardness. Unfortunately they are rather brittle, lacking toughness, when very hard. Where medium hardness with reasonable toughness is required, a carbon steel with about 0.7% carbon can be used. Where hardness is the primary consideration and toughness is not important, a carbon steel with about 1.2% carbon can be used.

Alloy tool steels are made harder and more wear resistant by the addition to the steel of elements that promote the production of stable hard carbides. Manganese, chromium, molybdenum, tungsten and vanadium are examples of such elements. A manganese tool steel contains from about 0.7 to 1.0% carbon and 1.0 to 2.0% manganese. Such a steel is oil quenched from about 780–800°C and then tempered. The manganese content may be partially replaced by chromium, such a change improving the toughness. Shock-resistant tool steels are designed to have toughness under impact conditions. For such properties, fine grain is necessary and this is achieved by the addition of vanadium. Tool steels designed for use with hot-working processes need to maintain their properties at the temperatures used. Chromium and tungsten, when added to steels, form carbides which are both stable and hard, hence maintaining the properties to high temperatures. Steels used at high machining speeds are called high speed tool steels. The high speed results in the material becoming hot. Such steels must not be tempered by the high temperatures produced during machining. The combination of tungsten and chromium is found to give the required properties, the carbides formed by these elements being particularly stable at high temperatures.

See Coding of tool steels, Composition of tool steels, Tool steel properties and Uses of tool steels.

3.2 Codes and compositions

Coding system for carbon steels

See Coding system for steels.

Coding system for cast irons

The codes given to cast irons tend to relate to their mechanical properties. Thus grey cast irons, according to British Standards (BS), are designated in seven grades: 150, 180, 220, 260, 300, 350 and 400. These numbers are the minimum tensile strengths in MPa or MN m^{-2} or N mm^{-2} in a 30 mm diameter test bar. The American standards, American Society for Testing Materials (ASTM), use the minimum strengths in k.s.i. (klb in^{-2}).

Malleable cast irons are specified, according to British Standards, by a letter B, P or A to indicate whether blackheart, pearlitic or whiteheart, followed by a number to indicate the minimum requirements for section sizes greater than 15 mm of the tensile strength in MPa or MN m^{-2} or N mm^{-2} and a number to indicate the percentage elongation. Thus, for example, B340/12 is a blackheart cast iron with a minimum tensile strength of 340 MPa and a percentage elongation of 12%. In American standards, the specification of malleable cast irons is in terms of two numbers to indicate the yield stress in 10^2 p.s.i. followed by the percentage elongation.

Ductile irons, according to British Standards, are specified by two numbers, the first representing the minimum tensile strength in MPa or MN m^{-2} or N mm^{-2} and the second the percentage elongation, e.g. 420/12. In American standards, ductile irons are specified by three numbers, the first representing the minimum tensile strength in k.s.i, the second the yield stress in k.s.i., the third the percentage elongation, e.g. 60–40–18.

Alloy cast irons, according to British Standards, are specified for graphite-free white irons by a number, 1, 2 or 3, to indicate the type of alloy (1 is low alloy, 2 is nickel–chromium, 3 is high chromium), followed by a letter to indicate the specific alloy. For graphite containing irons, austenitic irons containing flaky graphite are designated by the letter L and spheroidal graphite by S. This is followed by numbers and letters indicating the percentages of the main alloying elements. Ferritic high silicon, graphite containing alloys are designated by the main alloying elements and their percentages, e.g. Si10.

Coding system for stainless steels

The British Standard system of coding these steels is linked to the American Iron and Steel Institute (AISI) system. In general, the first three digits of the British system relate to the three digits used for the AISI system. The British system then follows these three digits by the letter S and then a further two digits in the range 11 to 99. These indicate variants of the steel specified by the first three digits. The range of numbers used for the first three digits, on the British system, is 300 to 499. See Composition of stainless steels for direct equivalents.

Coding system for steels: American

The AISI–SAE system (American Iron and Steel Institute, Society of Automotive Engineers) uses a four-digit code. The first two numbers indicate the type of steel, with the first digit indicating the grouping by major alloying element and the second digit in some instances indicating the approximate percentage of that element. The third and fourth numbers are used to indicate 100 times the percentage of carbon content. Table 3.2 outlines the AISI-SAE code system.

Table 3.2 AISI-SAE code system

Number series	Form of steel	Examples of series subdivision
1000	Carbon steel	10XX Plain carbon with maximum of 1% manganese
		11XX Resulphurized
		12XX Resulphurized and rephosphorized

	Manganese steels	*15XX* Plain carbon with 1.00 to 1.65% manganese *13XX* has 1.75% manganese
2000	Nickel steels	*23XX* has 3.5% nickel *25XX* has 5.0% nickel
3000	Nickel-chromium	*31XX* has 1.25% nickel, 0.65% or 0.80% chromium *32XX* has 1.75% nickel, 1.07% chromium *33XX* has 3.50% nickel, 1.50% or 1.57% chromium *34XX* has 3.00% nickel, 0.77% chromium
4000	Molybdenum steels	*40XX* has 0.20 or 0.25% molybdenum *44XX* has 0.40 or 0.52% molybdenum
	Chromium-molybdenum steels	*41XX* has 0.50%, 0.80% or 0.95% chromium, 0.12%, 0.20%, 0.25% or 0.30% molybdenum
	Nickel-chromium-molybdenum steels	*43XX* has 1.82% nickel, 0.50% or 0.80% chromium, 0.25% molybdenum *S47XX* has 1.05% nickel, 0.45% chromium, 0.20% or 0.35% molybdenum
	Nickel-molybdenum steels	*46XX* has 0.85% or 1.82% nickel, 0.20% or 0.25% molybdenum
5000	Chromium steels	*50XX* has 0.27%, 0.40%, 0.50% or 0.65% chromium *51XX* has 0.80%, 0.87%, 0.92%, 0.95%, 1.00% or 1.05% chromium
6000	Chromium-vanadium steels	*61XX* has 0.60%, 0.80% or 0.95% chromium, 0.10% or 0.15% min. vanadium
7000	Tungsten-chromium steels	*72XX* has 1.75% tungsten, 0.75% chromium
8000	Nickel-chromium-molybdenum steels	*81XX* has 0.30% nickel, 0.40% chromium, 0.12% molybdenum *86XX* has 0.55% nickel, 0.50% chromium, 0.20% molybdenum *87XX* has 0.55% nickel, 0.50% chromium, 0.25% molybdenum *88XX* has 0.55% nickel, 0.50% chromium, 0.35% molybdenum
9000	Silicon-manganese steels	*92XX* has 1.40% or 2.00% silicon, 0.65%, 0.82% or 0.85% manganese, 0.00% or 0.65% chromium
	Nickel-chromium-molybdenum	*93XX* has 3.25% nickel, 1.20% chromium, 0.12% molybdenum

94XX has 0.45% nickel, 0.40% chromium, 0.12% molybdenum
97XX has 0.55% nickel, 0.20% chromium, 0.20% molybdenum
98XX has 1.00% nickel, 0.80% chromium, 0.25% molybdenum

To illustrate the use of the code in Table 3.2, consider a steel 1040. The first digit is 1 and so the steel is a carbon steel. The 10 indicates that it is a plain carbon steel with a maximum content of 1.00% manganese. The last pair of digits is 40 and so the steel contains 0.40% carbon.

There are also some specifically specified steels which have modified SAE numbers, a letter being included between the first pair and second pair of numbers (Table 3.3).

Table 3.3 Additional letters in SAE code

Code	Significance of letter
XXBXX	The B denotes boron intensified steels.
XXLXX	The L denotes leaded steels.

In addition to the four SAE digits, various letter prefixes and suffixes are given to provide additional information (Table 3.4).

Table 3.4 Letter prefixes and suffixes for SAE code

Prefix	Significance of prefix
A	Alloy steel made in an acid-hearth furnace
B	Carbon steel made in a Bessemer furnace
C	Carbon steel made in a basic open-hearth furnace
D	Carbon steel made in an acid open-hearth furnace
E	Made in an electric furnace
X	Composition varies from normal limits

Suffix	Significance of suffix
H	Steel will reach hardenability criteria

The American Society for Testing Materials (ASTM) and the American Society of Mechanical Engineers (ASME) also issue standards covering steels. Many of their grades are based on the AISI–SAE grades. Also see Coding system for stainless steels.

Coding system for steels: British

In Great Britain the standard codes for the specification of steels are specified by the British Standards Institution. The following codes are for wrought steels:

(1) The first three digits of the code designate the type of steel.

000 to 199	Carbon and carbon–manganese types, the number being 100 times the manganese content.
200 to 240	Free-cutting steels, the second and third numbers being approximately 100 times the mean sulphur content.
250	Silicon–manganese spring steels

300 to 499 Stainless and heat resistant valve steels
500 to 999 Alloy steels

(2) The fourth symbol is a letter.

A The steel is supplied to a chemical composition determined by chemical analysis.
H The steel is supplied to hardenability specification.
M The steel is supplied to mechanical property specification.
S The steel is stainless.

(3) The fifth and sixth digits correspond to 100 times the mean percentage carbon content of the steel.

To illustrate the above code system, consider a steel with the code 070M20. The first three digits 070 are between 000 and 199 and so the steel is a carbon or carbon–manganese type. The 070 indicates that the steel has 0.70% manganese. The fourth digit is M and so the steel is supplied to mechanical property specification. The fifth and sixth digits are 20 and so the steel has 0.20% carbon.

The first three digits used for alloy steels are subdivided, as shown in Table 3.5, according to the main alloying elements.

Table 3.5 Coding system for BS alloy steels

Digits	Main alloying elements
500–519	Ni
520–539	Cr
540–549	Mo
550–569	V, Ti, Al, Nb
570–579	Si–Ni, Si–Cr, Si–Mo, Si–V
580–589	Mn–Si, Mn–Ni
590–599	Mn–Cr
600–609	Mn–Mo
610–619	Mn–V
620–629	Ni–Si, Ni–Mn
630–659	Ni–Cr
660–669	Ni–Mo
670–679	Ni–V, Ni–X
680–689	Cr–Si, Cr–Mn
690–699	Cr–Ni
700–729	Cr–Mo
730–739	Cr–V
740–749	Cr–X
750–759	Mo–Cr, Mo–V, Mo–X
760–769	Si–Mn–Cr
770–779	Mn–Ni–Cr
780–789	Mn–Ni–Mo
790–799	—
800–839	Ni–Cr–Mo
840–849	Ni–Cr–V, Ni–Cr–X
850–859	Ni–Mo–V, Ni–Cr–X
860–869	—
870–879	Cr–Ni–Mo
880–889	—
890–899	Cr–Mo–V
900–909	Cr–Al–Mo
910–919	—
920–929	Si–Mn–Cr–Mo
930–939	—
940–949	Mn–Ni–Cr–Mo

950–969	–
970–979	Ni–Cr–Mo–V
980–999	

Note: In the above table, where elements are separated by a dash, e.g. Mo–Cr, then those are all present as main alloying elements. Where elements, or groups of elements, are separated by a comma, e.g. Mo–Cr, Mo–V, Mo–X, then each alone has a number in the specified range of digits. Where the symbol X is specified then this means some element other than those already specified.

To give an illustration from Table 3.5, 805M20 is a Ni–Cr–Mo alloy since the first three digits fall within the 800–839 category.

Steels are frequently specified to mechanical properties. This is usually according to the tensile strength in the hardened and tempered condition. A code letter can be used to indicate the tensile strength range in which a steel falls when in this condition. The letter is said to refer to the condition of the steel. Table 3.6 gives the condition codes.

Table 3.6 Condition codes
Condition code   Tensile strength range (MPa or MN m^{-2})

P	550 to 700
Q	629 to 770
R	700 to 850
S	770 to 930
T	850 to 1000
U	930 to 1080
V	1000 to 1150
W	1080 to 1240
X	1150 to 1300
Y	1240 to 1400
Z	1540 minimum

Also see Coding system for stainless steels.

Coding system for tool steels

The British coding for tool steels is based on that of the AISI, the only difference being that in the British code the American codes are prefixed by the letter B (Table 3.7).

Table 3.7 Coding system for tool steels
AISI	BS	Material
W	BW	A water-hardening tool steel
O	BO	Oil-hardening tool steel for cold work
A	BA	Medium alloy hardening for cold work
D	BD	High-carbon and high-chromium content for cold work
H	BH	Chromium or tungsten base for hot work
M	BM	Molybdenum base, high speed steel
T	BT	Tungsten base, high speed steel
S	BS	Shock-resisting tool steel
P	BP	Mould steels
L	BL	Low alloy tool steel for special applications
F	BF	Carbon–tungsten steels

Note: The tool steel code letters are followed by a number to denote a particular steel composition.

Composition of alloy steels

Tables 3.8 and 3.9 give the compositions of commonly used alloy steels. Table 3.8 gives compositions in relation to AISI–SAE alloy steel specification, BS equivalents being given where appropriate. Table 3.9 gives compositions of BS steels, AISI equivalents being given where appropriate.

Table 3.8 AISI–SAE composition of commonly used alloy steels

AISI	BS	Mean percentages					
		C	Mn	Cr	Mo	Ni	Other
Manganese steels							
1330		0.30	1.75				
1340		0.40	1.75				
Molybdenum steels							
4037		0.38	0.80		0.25		
Chromium–molybdenum steels							
4130		0.30	0.50	0.95	0.20		
4140	708/9M40	0.40	0.88	0.95	0.20		
Nickel–chromium–molybdenum steels							
4320		0.20	0.55	0.50	0.25	1.83	
4340	817M40	0.40	0.70	0.80	0.25	1.83	
Nickel–molybdenum steels							
4620		0.19	0.53		0.25	1.83	
4820		0.19	0.60		0.25	3.50	
Chromium steels							
5130	530A30	0.30	0.80	0.95			
5140	530M40	0.40	0.80	0.80			
5160		0.60	0.88	0.80			
Chromium–vanadium steels							
6150	735A50	0.50	0.80	0.95			0.15 V
Nickel–chromium–molybdenum steels							
8620	805M20	0.20	0.75	0.50	0.20	0.55	
8630		0.30	0.75	0.50	0.20	0.55	
8640		0.40	0.88	0.50	0.20	0.55	
8650		0.50	0.88	0.50	0.20	0.55	
8660		0.60	0.88	0.50	0.20	0.55	
Silicon steels							
9260	250A58	0.60	0.88				2.0 Si

Note: Most of the steels contain about 0.15 to 0.30 Si, and less than 0.035 P and 0.040 S.

Table 3.9 BS Composition of alloy steels

BS	AISI	Mean percentages					
		C	Mn	Cr	Mo	Ni	Other
Manganese steels							
120M19		0.19	1.20				
120M28		0.28	1.20				
120M36	1039	0.36	1.20				
150M19		0.19	1.50				
150M28		0.28	1.50				
150M36	1039	0.36	1.50				
Silicon manganese steels							
250A53		0.53	0.85				Si 1.9
250A58	9260	0.58	0.85				Si 1.9

Chromium steels

526M60	0.60	0.65	0.65		
530M40 5140	0.40	0.75	1.05		

Manganese-molybdenum steels

605M36	0.36	1.50		0.27		
606M36	0.36	1.50		0.27		P <0.06 S 0.15-0.25
608M38	0.38	1.50		0.50		

Nickel-chromium steels

653M31	0.31	0.60	1.00		3.00

Chromium-molybdenum steels

708M40 4137/40	0.40	0.70	0.90	0.20	
709M40 4140	0.40	0.60	0.90	0.30	<0.40
722M24	0.24	0.45	3.00	0.55	

Nickel-chromium-molybdenum steels

817M40 4340	0.40	0.55	1.20	0.30	1.50
826M40	0.40	0.55	0.65	0.50	2.55
835M30	0.30	0.55	1.25	0.27	4.10

Manganese-nickel-chromium-molybdenum steels

945M38	0.38	1.40	0.50	0.20	0.75

Composition of carbon steels

Tables 3.10 and 3.11 give the compositions of commonly used carbon steels. Table 3.10 gives the AISI compositions, with BS equivalents where appropriate. Table 3.11 gives the BS compositions, with AISI equivalents where appropriate.

Table 3.10 AISI-SAE composition of common carbon steels

		Percentages	
AISI	BS	C	Mn
1006		0.08 max	0.25-0.40
1010		0.08-0.13	0.30-0.60
1015		0.13-0.18	0.30-0.60
1020	070M20	0.18-0.23	0.30-0.60
1025		0.22-0.28	0.30-0.60
1030		0.28-0.34	0.60-0.90
1035	080M36	0.32-0.38	0.60-0.90
1040	080A40	0.37-0.44	0.60-0.90
1045		0.43-0.50	0.60-0.90
1050		0.48-0.55	0.60-0.90
1055		0.50-0.60	0.60-0.90
1060	060A62	0.55-0.66	0.60-0.90
1065		0.60-0.70	0.60-0.90
1070		0.65-0.75	0.60-0.90
1075		0.70-0.80	0.40-0.70
1080		0.75-0.88	0.60-0.90
1085		0.80-0.93	0.70-1.00
1090		0.85-0.98	0.60-0.90
1095	060A96	0.90-1.03	0.30-0.50

Note: There is a maximum of 0.040% phosphorus and 0.05% sulphur.

Table 3.11 BS composition of common carbon steels

BS	AISI	Mean percentages	
		C	Mn
070M20	1020	0.20	0.70
070M26		0.26	0.70
080M30		0.30	0.80
080M36	1035	0.36	0.80
080M40	1043	0.40	0.80
080M46	1043	0.46	0.80
080M50		0.50	0.80
070M55		0.55	0.70

Composition of cast irons

Table 3.12 shows, in general terms, the compositions of the various forms of unalloyed cast iron.

Table 3.12 Composition ranges of unalloyed cast irons

Cast iron	Percentages				
	C	Si	Mn	S	P
Grey	2.5–4.0	1.0–3.0	0.25–1.00	0.02–0.25	0.05–1.00
Ductile	3.0–4.0	1.8–2.8	0.10–1.00	0.03 max.	0.10 max.
White	1.8–3.6	0.5–1.9	0.25–0.80	0.06–0.20	0.06–0.18
Malleable	2.0–2.6	1.1–1.6	0.20–1.00	0.04–0.18	0.18 max.

The addition of elements, such as nickel, chromium, or molybdenum to unalloyed white cast iron, can change its pearlitic structure to martensite, bainite or austenite to give what is often termed abrasion-resistant white irons. Corrosion-resistant irons are produced if high percentages of silicon, or chromium or nickel are used. Heat-resistant grey and ductile cast irons are produced if silicon, chromium, nickel, molybdenum or aluminium are added. Table 3.13 shows the compositions of a selection of these alloy cast irons.

Table 3.13 Composition of alloy cast irons

Material	Composition (%)
Abrasion-resistant white	
BS low alloy, 1A, 1B, 1C	C 2.4–3.4, Si 0.5–1.5, Mn 0.2–0.8, Cr ⟨2.0.
BS nickel–chromium, 2A–E	C 2.7–3.6, Si 0.3–2.2, Mn 0.2–0.6, Ni 3.0–6.0, Cr 1.5–10.0.
BS high chromium, 3A–E	C 2.4–3.2, Si ⟨1.0, Mn 0.5–1.5, Cr 14.0–17.0, Mo ⟨3.0, Ni ⟨1.0, Cu ⟨1.2.
Martensitic nickel–chromium	C 3.00–3.60, Si 0.40–0.70, Mn 0.40–0.70, Cr 1.40–3.50, Ni 4.00–4.75, P ⟨0.40, S ⟨0.15.
High chromium white	C 2.25–2.85, Si 0.25–1.00, Mn 0.50–1.25, Cr 24.0–30.0, P ⟨0.40, S ⟨0.15.
Ni hard, 3.5 Ni Cr	C 2.8–3.6, Ni 2.5–4.75, Cr 1.2–1.35, Si 0.4–0.7, Mn

0.2-0.7

Corrosion-resistant

BS Ferritic high silicon Si 10	Si 10
BS Ferritic high silicon Si 14	Si 14
BS Ferritic high silicon Si Cr 14 4	Si 14, Cr 4
BS Ferritic high silicon Si 16	Si 16
High silicon	C 0.4-1.1, Si 14-17, Mn <1.5, Cr <5.0, Mo <1.0, Cu <0.5, P <0.15, S <0.15
High chromium	C 1.2-4.0, Si 0.5-3.0, Mn 0.3-1.5, Cr 12-35, Mo <4.0, Cu <3.0, Ni <5.0, P <0.15, S <0.15
Ni-resistant austenitic	C <3.0, Si 1.0-2.8, Mn 0.5-1.5, Cr 1.5-6.0, Mo <1.0, Cu <7.0, Ni 13.5-36, P <0.08, S <0.12

Heat-resistant grey

BS Austenitic L-Ni Mn 13 7	C <3.0, Ni 12-14, Mn 6.0-7.0, Si 1.5-3.0
BS Austenitic L-Ni Cu Cr 15 6 2	C <3.0, Ni 13.5-17.5, Cu 5.5-7.5, Cr 1.0-2.5, Si 2.0-2.8, Mn 1.0-1.5.
BS Austenitic L-Ni Cu Cr 15 6 3	C <3.0, Ni 13.5-17.5, Cu 5.5-7.5, Cr 2.5-3.5, Si 1.0-2.8, Mn 1.0-1.5
BS Austenitic L-Ni Cr 20 2	C <3.0, Ni 18-22, Cr 1.0-2.5, Si 1.0-2.8, Mn 1.0-1.5
BS Austenitic L-Ni Cr 20 3	C <3.0, Ni 18-22, Cr 2.5-3.5, Si 1.0-2.8, Mn 1.0-1.5
BS Austenitic L-Ni Si Cr 30 5 5	Ni 30, Si 5, Cr 5
BS Austenitic L-Ni 35	C <2.4, Ni 34-36, Si 1.0-2.0, Mn 0.4-0.8
Medium silicon	C 1.6-2.5, Mn 0.4-0.8, Si 4.0-7.0, P <0.30, S <0.10.
High chromium	C 1.8-3.0, Mn 0.3-1.5, Si 0.5-2.5, Ni <5.0, Cr 15-35, P <0.15, S <0.15
Nickel-chromium	C 1.8-3.0, Mn 0.4-1.5, Si 1.0-2.75, Ni 13.5-36, Cr 1.8-6.0, Mo <1.0, Cu <7.0, P <0.15, S <0.15
Nickel-chromium-silicon	C 1.8-2.6, Mn 0.4-1.0, Si 5.0-6.0, Ni 13-43, Cr 1.8-5.5, Mo <1.0, Cu <10.0, P <0.10, S <0.10
High aluminium	C 1.3-2.0, Mn 0.4-1.0, Si 1.3-6.0, Al 20-25, P <0.15, S <0.15

Heat-resistant ductile

BS Austenitic S-Ni Mn 13 7	C <3.0, Ni 12-14, Mn 6-7, Si 2-3
BS Austenitic S-Ni Cr 20 2	C <3.0, Ni 18-22, Cr 1.0-2.5, Si 1.0-2.8, Mn 0.7-1.5, P <0.08
BS Austenitic S-Ni Cr 20 3	C <3.0, Ni 18-22, Cr 2.5-3.5, Si 1.0-2.8, Mn 0.7-1.5, P <0.08
BS Austenitic S-Ni Si Cr 20 5 2	C <3.0, Ni 18-22, Si 4.5-5.5, Cr 1.0-2.5, Mn 1.0-1.5, P

	<0.08
BS Austenitic S-Ni 22	C <3.0, Ni 21-24, Si 1.0-2.8, Mn 1.8-2.4, P <0.08
BS Austenitic S-Ni Mn 23 4	C <2.6, Ni 22-24, Mn 4.0-4.4, Si 1.9-2.6
BS Austenitic S-Ni Cr 30 1	Ni 30, Cr 1
BS Austenitic S-Ni Cr 30 3	C <2.6, Ni 28-32, Cr 2.5-3.5, Si 1.5-2.8, Mn <0.5, P <0.08
BS Austenitic S-Ni Si Cr 30 5 5	Ni 30, Si 5, Cr 5
BS Austenitic S-Ni 35	C <2.4, Ni 34-36, Si 1.5-2.8, Mn <0.5
BS Austenitic S-Ni Cr 35 3	C <2.4, Ni 34-36, Cr 2-3, Si 1.5-2.8, Mn <0.5
Medium silicon	C 2.8-3.8, Mn 0.2-0.6, Si 2.5-6.0, Ni <1.5, P <0.08, S <0.12
Nickel-chromium	C <3.0, Mn 0.7-2.4, Ni 18-36, Si 1.75-5.5, Cr 1.75-3.5, Mo <1.0, P <0.08, S <0.12

Note: The carbon contents indicated are the total carbon contents and not just the free carbon.

Composition of free-cutting steels

Free-cutting steels are often referred to as resulphurized steels since sulphur is the main element added to give the free-cutting properties. Tables 3.14 and 3.15 give the composition of commonly used free-cutting steels.

Table 3.14 Composition of AISI free-cutting steels

AISI	Percentages		
	C	Mn	S
1110	0.08-0.13	0.30-0.60	0.08-0.13
1117	0.14-0.20	1.00-1.30	0.08-0.13
1118	0.14-0.20	1.30-1.60	0.08-0.13
1137	0.32-0.43	1.35-1.65	0.13-0.20
1140	0.37-0.44	0.70-1.00	0.08-0.13
1141	0.37-0.45	1.35-1.65	0.08-0.13
1144	0.40-0.48	1.35-1.65	0.24-0.33
1146	0.42-0.49	0.70-1.00	0.08-0.13
1151	0.48-0.55	0.70-1.00	0.08-0.13

Table 3.15 Composition of BS free-cutting steels

BS	Percentages		
	C	Mn	S
210M15	0.12-0.18	0.90-1.30	0.10-0.18
212M36	0.32-0.40	1.00-1.40	0.12-0.20
214M15	0.12-0.18	1.20-1.60	0.10-0.18
216M36	0.32-0.40	1.30-1.70	0.12-0.20
216M44	0.40-0.48	1.20-1.50	0.12-0.20
220M07	0.15	0.90-1.30	0.20-0.30
226M44	0.40-0.48	1.30-1.70	0.22-0.30
230M07	0.15	0.90-1.30	0.20-0.30

Composition of maraging steels

Table 3.16 shows the compositions of typical maraging steels. See Maraging steels.

Table 3.16 Composition of maraging steels

Grade	Percentages					
	Ni	Co	Mo	Al	Ti	C (max.)
200	18	8	3.2	0.1	0.2	0.03
250	18	8	5.0	0.1	0.4	0.03
300	18	9	5.0	0.1	0.6	0.03
350	18	12	4.0	0.1	1.8	0.01

Composition of stainless steels

Tables 3.17 and 3.18 give the compositions of typical stainless and heat resisting steels. The stainless steels owe their high resistance to corrosion to the inclusion of higher percentages of chromium in the composition of the alloys. A minimum of 12% chromium is required. Stainless steels are divided into three main categories according to their microstructure, these being austenitic, ferritic and martensitic. Table 3.17 gives the steels in terms of AISI codes, with equivalent BS steels being indicated where appropriate. Table 3.18 gives BS codes with equivalent AISI codes being indicated.

Table 3.17 Composition of AISI stainless steels

AISI	BS	Percentages				
		C	Cr	Ni	Mn	Other
Austenitic						
201		0.15	17	4.5	6.0	
301		0.15	17	7	2.0	
302	302S31	0.15	18	9	2.0	
304	304S15	0.08	19	9.5	2.0	
	304S16, S18, S25, S40					
309		0.20	23	13.5	2.0	
316	316S16	0.08	17	12	2.0	
	316S18, S25, S26, S30, S33, S40, S41					
321	321S12	0.08	18	10.5	2.0	Ti 5 × %C
	321S18, S22, S27, S31, S40, S49, S50, S59, S87					
Ferritic						
405	405S17	0.08	13		1.0	Al 0.20
409		0.08	11		1.0	Ti 6 × %C
430	430S15	0.12	17		1.0	
442		0.20	20.5		1.0	
446		0.20	25.0		1.5	
Martensitic						
403	420S29	0.15	12.2		1.0	
410	410S21	0.15	12.5		1.0	
	410S27					
420	420S37	>0.15	12		1.0	
431	431S29	0.20	16	1.8	1.0	
440C		1.07	17		1.0	

Note: There is also generally a maximum of 0.030% S, 0.040% P and 1.00% Si.

Table 3.18 Composition of BS stainless steels

BS	AISI	Percentage				
		C	Cr	Ni	Mn	Other
Austenitic						
302S31	302	0.12	18	9	2.0	
303S31	303	0.12	18	9	2.0	
304S31	304	0.07	18	9.5	2.0	
310S31		0.15	25	20.5	2.0	
316S33	316	0.07	17	12.5	2.0	2.8 Mo
321S31	321	0.08	18	10.5	2.0	5C Ti
Ferritic						
403S17		0.08	13	0.5	1.0	
430S17	430	0.08	17	0.5	1.0	
Martensitic						
410S21	410	0.12	12.5	1.0	1.0	
416S21	416	0.12	12.5	1.0	1.5	
420S29	403	0.17	12.5	1.0	1.0	
420S37	420	0.24	13	1.0	1.0	
431S29	431	0.16	16.5	2.5	1.0	

Note: Generally all the steels contain about 0.05% P, 0.03% S and 1.0% Si.

Composition of tool steels

Table 3.19 gives the composition of commonly used tool steels. In addition to steel, sintered carbides (see Chapter 10), ceramics (see Chapter 10) and cobalt–chromium–tungsten–molybdenum alloys are used. The identification codes used are those of the AISI; the BS codes are the same but just preceded by the letter B.

Table 3.19 Composition of tool steels

AISI	Percentage					
	C	W	Mo	Cr	V	Other
Water-hardening						
W1	0.6–1.4					
W2	0.6–1.4				0.25	
W5	1.10			0.50		
Shock-resisting						
S1	0.50	2.50		1.50		
S2	0.50		0.50			1.00 Si
S5	0.55		0.40			0.80 Mn, 2.00 Si
S7	0.50		1.40	3.25		
Oil-hardening						
O1	0.90	0.50		0.50		1.00 Mn
O2	0.90					1.60 Mn
O6	1.45		0.25			0.80 Mn, 1.00 Si
O7	1.20			1.75	0.75	
Air-hardening						
A2	1.00		1.00	5.00		
A3	1.25		1.00	5.00	1.00	
A4	1.00		1.00	1.00		2.00 Mn
A6	0.70		1.25	1.00		2.00 Mn
A7	2.25	1.00	1.00	5.25	4.75	W optional
A8	0.55	1.25	1.25	5.00		
A9	0.50		1.40	5.00	1.00	1.50 Ni
A10	1.35		1.50			1.80 Mn, 1.25 Si, 1.80 Ni

	C	W	Mo	Cr	V	Co
High carbon-high chromium steels						
D2	1.50		1.00	12.00	1.00	
D3	2.25			12.00		
D4	2.25		1.00	12.00		
D5	1.50		1.00	12.00		3.00 Co
D7	2.35		1.00	12.00	4.00	
Hot-work (chromium) steels						
H10	0.40		2.50	3.25	0.40	
H11	0.35		1.50	5.00	0.40	
H12	0.35	1.50	1.50	5.00	0.40	
H13	0.35		1.50	5.00	1.00	
H14	0.40	5.00		5.00		
H19	0.40	4.25		4.25	2.00	4.25 Co
Hot-work (tungsten) steels						
H21	0.35	9.00		3.50		
H22	0.35	11.00		2.00		
H23	0.30	12.00		12.00		
H24	0.45	15.00		3.00		
H25	0.25	15.00		4.00		
H26	0.50	18.00		4.00	1.00	
Hot-work molybdenum						
H42	0.60	6.00	5.00	4.00	2.00	
High-speed tungsten						
T1	0.75	18.00		4.00	1.00	
T2	0.80	18.00		4.00	2.00	
T4	0.75	18.00		4.00	1.00	5.00 Co
T5	0.80	18.00		4.00	2.00	8.00 Co
T6	0.80	20.00		4.50	1.50	12.00 Co
High-speed molybdenum						
M1	0.85	1.50	8.50	4.00	1.00	
M2	0.85/1.00	6.00	5.00	4.00	2.00	
M4	1.30	5.50	4.50	4.00	4.00	
M6	0.80	4.00	5.00	4.00	1.50	12.00 Co
M7	1.00	1.75	8.75	4.00	2.00	
M10	0.85/1.00		8.00	4.00	2.00	
M30	0.80	2.00	8.00	4.00	1.25	5.00 Co
M33	0.90	1.50	9.50	4.00	1.15	8.00 Co
M34	0.90	2.00	8.00	4.00	2.00	8.00 Co
M36	0.80	6.00	5.00	4.00	2.00	8.00 Co
M41	1.10	6.75	3.75	4.25	2.00	5.00 Co
M42	1.10	1.50	9.50	3.75	1.15	8.00 Co
M43	1.20	2.75	8.00	3.75	1.60	8.25 Co
M44	1.15	5.25	6.25	4.25	2.00	12.00 Co
M46	1.25	2.00	8.25	4.00	3.20	8.25 Co
M47	1.10	1.50	9.50	3.75	1.25	5.00 Co

3.3 Heat treatment

Annealing

Tables 3.20 and 3.21 give the annealing temperatures appropriate for a range of AISI and BS carbon and alloy steels. Table 3.22 gives the annealing temperatures for stainless steels. Data specific to particular steels are given, together with the consequential properties, in the section Mechanical properties.

Table 3.20 Annealing temperatures for AISI carbon and alloy steels

AISI	Annealing temperature °C
Carbon steels	
1020	855–900
1025	855–900
1030	845–885
1035	845–885
1040	790–970
1045	790–870
1050	790–870
1060	790–845
1070	790–845
1080	790–845
1090	790–830
1095	790–830
Alloy steels	
1330	845–900
1340	845–900
4037	815–855
4130	790–845
4140	790–845
4340	790–845
5130	790–845
5140	815–870
5160	815–870
6150	845–900
8630	790–845
8640	815–870
8650	815–870
8660	815–870
9260	815–870

Table 3.21 Annealing temperatures for BS carbon and alloy steels

BS	Annealing temperature (°C)
Carbon steels	
070M20	880–910
080M30	860–890
080M40	830–860
080M50	810–840
070M55	810–840
Alloy steels	
120M36	840–870
150M19	840–900
150M36	840–870
530M40	810–830
605M36	830–860
606M36	830–860
708M40	850–880
817M40	820–850
826M40	820–850
835M30	810–830

Table 3.22 Annealing temperatures for stainless steels

Material	Annealing temperature (°C)
Austenitic	
201, 301, 302, 304	1010–1120
309, 316	1040–1120
321	955–1065
Ferritic	
405	650–815
409	870–900
430	705–790
Martensitic	
403, 410, 420	830–885
431	Not recommended
440C	845–900

Note: The coding used to specify the steels is AISI, for the equivalent BS codes see Coding system for stainless steels and Composition of stainless steels.

Case hardening

See Surface hardening for details of processes in relation to ferrous alloys.

Nitriding

See Surface hardening for details of processes in relation to ferrous alloys.

Surface hardening

The surface hardness of steels can be changed by surface-hardening treatments. Table 3.23 shows the effects of such treatments. Some steels are designed primarily for carburizing to give a surface with a high resistance to wear and a core with adequate strength and toughness for the required loading. Such steels are referred to as carburizing or case hardening steels, the following being examples of such, to British Standards: carbon steels 045M10, 080M15, 130M15, 210 M15; boron steels 170H15, 173H16, 174H20, 175H23; alloy steels 523H15, 527H17, 590H17, 635H15, 637H17, 655H13, 665H17, 665H20, 665H23, 708H20, 805H29, 805H22, 808H17, 815H17, 820H17, 822H17, 832H13, 835H15. Steels, to British Standards, suitable for nitriding include: 708M40, 709M40, 720M32, 722M24, 897M39, 905M39.

Table 3.23 Effects of surface-hardening treatments

Process	Temp. (°C)	Case		Main materials
		Depth (mm)	Hardness (HRC)	
Pack carburizing	810–1100	0.25–3	45–65	Low carbon and carburizing alloy steels.
Gas carburizing	810–980	0.07–3	45–65	Low carbon and carburizing alloy steels.
Cyaniding	760–870	0.02–0.7	50–60	Low carbon and alloy steels
Nitriding	500–530	0.07–0.7	50–70	Alloy steels.
Carbo-nitriding	700–900	0.02–0.7	50–60	Low carbon and alloy steels.

| Flame hardening | 850–1000 | Up to 0.8 | 55–65 | 0.4 to 0.7% carbon steels. |
| Induction hardening | 850–1000 | 0.5–5 | 55–65 | 0.4 to 0.7% carbon steels. |

Tempering

Tempering is the name given to the process in which a steel, hardened as a result of quenching, is reheated to a temperature below the A_1 critical temperature in order to modify the martensitic structure of the steel. After quenching the tensile strength is a maximum and the toughness a minimum. In general, the higher the tempering temperature the lower the resulting tensile strength and the higher the toughness. Figure 3.2 shows the type of changes in properties produced for an alloy steel (0.40% carbon, 0.70% manganese, 1.8% nickel, 0.80% chromium and 0.25% molybdenum).

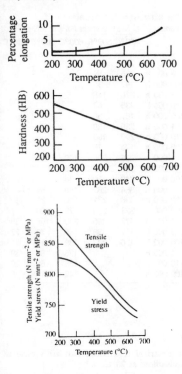

Figure 3.2 The effect of tempering on the properties of a steel

See the data given with Tensile properties for the effects of tempering on the properties of a range of steels.

3.4 Properties

Creep properties

The approximate high temperature limits of use of ferrous alloys, set by creep (rupture stress falling below about 50 MPa or MN m^{-2} in 100 000 hours) are given in Table 3.24. Table 3.25 gives values of rupture stresses at different temperatures for some typical steels.

Table 3.24 High temperature limits of ferrous alloys

Material	High temp. limit
Cast irons	350°C
Carbon steels	450°C
Chromium–molybdenum steels	550°C
18% chromium–10% nickel steel	600°C
25% chromium–20% nickel steel	750°C

Table 3.25 Rupture stresses for steels

Composition (% main elements) Heat treated condition		Rupture stress (MPa or MN m^{-2})				
		400°C	500°C	600°C	700°C	800°C
0.2 C, 0.75 Mn	1 000 h	295	118			
Norm. 920°C,	10 000 h	225	59			
Temp. 600°C	100 000 h	147	30			
0.17 C, 1.3 Mn	1 000 h	310	118			
Norm. 920°C,	10 000 h	235	67			
Temp. 600°C	100 000 h	167	30			
5 Cr, 0.5 Mo	1 000 h	230	130			
Annealed	10 000 h	200	100			
	100 000 h	170	75			
9 Cr, 1 Mo	1 000 h	275	200	100		
Norm. 990°C,	10 000 h	260	170	75		
Temp. 750°C	100 000 h	240	130	54		
18 Cr, 8 Ni	1 000 h	354	262	146	54	
Air cool 1050°C	10 000 h	336	231	100	31	
	100 000 h	323	200	70	20	
18 Cr, 12 Ni, 2 Mo	1 000 h	477	385	230	108	35
Norm. 1050°C, air cool	10 000 h	462	338	185	70	22
	100 000 h	430	293	139	46	11
25 Cr, 20 Ni	1 000 h			175	88	39
1000/1150°C air cool/	10 000 h			135	58	20
quench	100 000 h			42	17	6

Density

See Specific gravity.

Electrical resistivity

Table 3.26 shows the electrical resistivities that are typical of different types of ferrous alloys at 20°C.

Table 3.26 Electrical resistivities of ferrous alloys

Material	Resistivity $\mu\Omega$ m
Mild steel	0.16
Medium carbon steel	0.17
Manganese steel	0.23
Free-cutting steel	0.17

Nickel–manganese steel	0.23–0.39
Chromium steels	0.22
Chromium–molybdenum steel	0.22
Nickel–chromium–molybdenum steel	0.25–0.27
Stainless, austenitic	0.69–0.78
Stainless, ferritic	0.60
Stainless, martensitic	0.55–0.70

Note: Resistivities are affected by the heat treatment that a metal has undergone.

Fatigue

The endurance limit for most ferrous materials is approximately 0.4 to 0.6 times the tensile strength. Values in practice do however very much depend on the surface conditions of the component concerned.

Hardness

For hardness values of steels see the tables given in this chapter under Mechanical properties or, in the case of tool steels, under Tool steels properties.

Impact properties

See the tables with Mechanical properties for values of impact energies for various steels. The values are given for temperatures of the order of 20°C. However, most steels become brittle at lower temperatures and this shows as a marked decrease in the impact energy of a material. For some steels this ductile-brittle transition occurs at about 0°C, for others it may be as low as –60°C or –80°C.

Machinability

The machinability of a steel is affected by its composition, its microstructure and its hardness. The machinability of an alloy is improved by the addition of sulphur and/or lead, whereas elements such as aluminium and silicon can reduce the machinability. Leaded and resulphurized steels are thus known as free-cutting steels because of their good machinability. The addition of sulphur and/or lead to an alloy does give some slight reduction in mechanical properties and welding is not generally recommended. Machinability is also affected by the microstructure. Thus, for example, ferritic steels have better machinability than martensitic steels.

A machinability index is used to give a measure of the ease of machining for materials. In the AISI system an index of 100% is specified for 1212 steel and all others are rated in relation to it. In the British Standards system the metal used is 070M20. 1212 is a resulphurized, rephosphorized steel while 070M20 is just a carbon steel. Tables 3.27 and 3.28 give typical values of this index for various ferrous materials.

Table 3.27 AISI machinability index values for steels

Material	AISI machinability index (%)
Carbon steels	
1010	55
1015	60
1020	65
1030	70
1040	60

1050	45
Resulphurized and rephosphorized carbon steels	
1211	94
1212	100
1213	136
Alloy steels	
1340	50
4130	70
4320	60
5140	65
6150	55
8640	65
8660	55

Table 3.28   BS machinability index values for steels

Material	BS machinability index (%)
Carbon Steels	
070M20	100
080M30	70
080M40	70
080M50	50
070M55	50
Alloy steels	
120M36	65
150M19	70
150M36	65
530M40	40
605M36	50
708M40	40
722M24	35
817M40	35
Free-cutting steels	
210M15	200
212M36	70
214M15	140
220M07	200

Mechanical properties of alloy steels

Tables 3.29 and 3.30 give the tensile properties of typical alloy steels, Table 3.29 according to American steel specifications and Table 3.30 to British Standards. The compositions of these steels can be seen in the section (in this chapter), Composition of alloy steels. Also included with the tensile properties are hardness and impact energies. The tensile modulus for all steels is about 200 to 207 GPa or GN m^{-2}.

Table 3.29   Mechanical properties of AISI-SAE alloy steels

AISI	Condition	Tensile strength (MPa)	Yield stress (MPa)	Elong- ation (%)	Hard- ness (BH)	Impact Izod (J)
Manganese steels						
1330	Q, T 200°C	1600	1450	9	460	
	Q, T 650°C	730	570	23	216	
1340	Q, T 200°C	1800	1600	11	500	
	Q, T 650°C	800	620	22	250	

N 870°C	840	560	22	250	90
A 800°C	700	440	26	200	70

Molybdenum steels

4037 Q, T 200°C	1030	760	6	310	
Q, T 650°C	700	420	29	220	

Chromium-molybdenum steels

4130 Q, T 200°C	1630	1460	10	470	
Q, T 650°C	810	700	22	245	
N 870°C	670	440	26	200	85
A 860°C	560	360	28	155	60
4140 Q, T 200°C	1770	1640	8	510	
Q, T 650°C	760	650	22	230	
N 870°C	1020	660	18	300	23
A 815°C	660	420	26	200	54

Nickel-chromium-molybdenum steels

4320 N 895°C	790	460	21	235	73
A 850°C	580	610	29	150	110
4340 Q, T 200°C	1880	1680	10	520	28
Q, T 650°C	970	860	19	280	50
N 870°C	1280	860	12	360	8
A 810°C	745	470	22	220	50

Nickel-molybdenum steels

4620 N 900°C	575	365	29	175	130
A 855°C	510	370	31	150	94
4820 N 860°C	750	485	24	230	110
A 815°C	680	465	22	200	93

Chromium steels

5130 Q, T 200°C	1610	1520	10	475	
Q, T 650°C	795	690	20	245	
5140 Q, T 200°C	1790	1640	9	490	
Q, T 650°C	760	660	25	235	
N 870°C	790	470	23	230	38
A 830°C	570	290	29	170	40
5160 Q, T 200°C	2200	1790	4	630	
Q, T 650°C	900	800	20	270	
N 855°C	960	530	18	270	11
A 812°C	720	275	17	200	10

Chromium-vanadium steels

6150 Q, T 200°C	1930	1690	8	540	
Q, T 650°C	945	840	17	280	
N 870°C	940	615	22	270	36
A 815°C	670	410	23	200	27

Nickel-chromium-molybdenum steels

8620 N 915°C	630	360	26	185	100
A 870°C	535	385	31	150	110
8630 Q, T 200°C	1640	1500	9	465	
Q, T 650°C	770	690	23	240	73
N 870°C	650	430	24	190	95
A 840°C	560	370	29	155	95
8640 Q, T 200°C	1860	1670	10	505	
Q, T 650°C	900	800	20	280	90
8650 Q, T 200°C	1940	1680	10	525	
Q, T 650°C	970	830	20	280	
N 870°C	1020	690	14	300	13
A 800°C	720	390	23	210	30
8660 Q, T 425°C	1630	1550	13	460	22
Q, T 650°C	965	830	20	280	81

Silicon steels

9260	Q, T 425°C	1780	1500	8	470
	Q, T 650°C	980	815	20	295

Note: Q = quenched, T = tempered, N = normalized, A = annealed.

Table 3.30 Mechanical properties of BS alloy steels

BS	Condition	Tensile strength (MPa)	Yield stress (MPa)	Elong-ation (%)	Hard-ness (BH)	Impact Izod (J)
Manganese steels						
120M19	Q, T	550–700	355	18	150–210	40
	N	460	265	19	140–190	34
120M28	Q, T	620–770	415	16	180–230	40
	N	530	330	16	150–210	34
120M36	Q, T	620–770	415	18	180–230	40
	N	570	340	16	170–220	
150M19	Q, T	550–700	340	18	150–210	54
	N	510	295	17	150–210	40
150M28	Q, T	620–770	400	16	180–230	47
	N	560	325	16	170–220	34
150M36	Q, T	620–770	400	18	180–230	47
	N	600	355	15	180–230	
Chromium steels						
526M60	Q, T	850–1000	620	11	250–300	
530M40	Q, T	700–850	525	17	200–260	54
Manganese–molybdenum steels						
605M36	Q, T	700–900	585	15	200–255	40
606M36	Q, T	700–850	525	15	200–255	40
608M38	Q, T	770–930	555	13	220–280	34
Nickel–chromium steels						
653M31	Q, T	770–930	585	15	220–280	20
Chromium–molybdenum steels						
708M40	Q, T	700–850	525	17	200–260	54
709M40	Q, T	770–930	555	13	220–280	27
722M24	Q, T	850–1000	650	13	250–300	27
Nickel–chromium–molybdenum steels						
817M40	Q, T	850–1000	650	13	250–300	40
826M40	Q, T	930–1250	740	12	270–330	34
835M30	Q, T	1540	1235	7	440	20
Manganese–nickel–chromium–molybdenum steels						
945M38	Q, T	700–850	495	15	200–260	34

Note: Q = quenched, T = tempered, N = normalized. All properties refer to ruling sections up to about 250 mm.

Mechanical properties of carbon steels

Tables 3.31 and 3.32 show the mechanical properties of carbon steels. The tensile modulus of all the steels can be taken as being about 200 to 207 GPa or GN m^{-2}. Figure 3.3 shows how the mechanical properties depend on the percentage of carbon in the alloy, and hence the microstructure present.

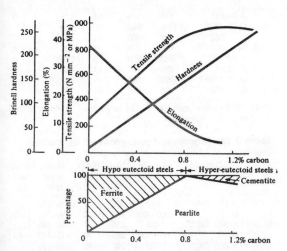

Figure 3.3 The effect of carbon content on the properties of carbon steels

Table 3.31 Mechanical properties of AISI carbon steels

AISI	Condition	Tensile strength (MPa)	Yield stress (MPa)	Elongation (%)	Hardness (BH)	Impact Izod (J)
1015	As rolled	420	315	39	125	110
	N 925°C	420	325	37	120	115
	A 870°C	385	285	37	110	115
1020	As rolled	450	330	36	145	85
	N 925°C	440	345	36	130	115
	A 870°C	395	295	37	110	120
1030	As rolled	550	345	32	180	75
	N 925°C	520	345	32	150	95
	A 840°C	465	340	31	125	70
	WQ, T 200°C	850	650	17	495	
	WQ, T 650°C	585	440	32	210	
1040	As rolled	620	415	25	200	50
	N 900°C	590	375	28	170	65
	A 790°C	520	355	30	150	45
	WQ, T 200°C	900	660	16	515	
	WQ, T 650°C	670	495	28	200	
1050	As rolled	725	415	20	230	30
	N 900°C	750	430	20	220	25
	A 790°C	635	365	24	185	16
	WQ, T 200°C	1120	810	9	515	
	WQ, T 650°C	720	540	28	235	
1060	As rolled	815	480	17	240	18
	N 900°C	775	420	18	230	13
	A 790°C	625	370	23	180	11
	Q, T 200°C	1100	780	13	320	
	Q, T 650°C	800	525	23	230	
1080	As rolled	965	585	12	290	7
	N 900°C	1010	525	11	290	7

A 790°C	615	375	25	175	6
Q, T 200°C	1310	980	12	390	
Q, T 650°C	890	600	21	255	
1095 As rolled	965	570	9	295	4
N 900°C	1015	500	10	295	5
A 790°C	655	380	13	190	3
Q, T 200°C	1290	830	10	400	
Q, T 650°C	895	550	21	270	

Note: WQ = water quenched, Q = quenched, N = normalized, A = annealed, T = tempered.

Table 3.32 Mechanical properties of BS carbon steels

BS	Condition	Tensile strength (MPa)	Yield stress (MPa)	Elongation (%)	Hardness (BH)	Impact Izod (J)
070M20	N 880–910°C	400	200	21	125–180	
	Q, T	550–700	355	20	150–210	40
070M26	N 870–900°C	430	215	20	140–190	
	Q, T	550–700	355	20	150–210	40
080M30	N 860–890°C	460	230	19	140–190	
	Q, T 550–660°C	550–700	340	18	150–210	40
080M36	N 840–870°C	490	245	18	140–190	
	Q, T 550–660°C	620–780	400	16	180–230	25
080M46	N 810–840°C	550	280	15	150–210	
	Q, T 550–660°C	625–775	370	16	180–230	
080M50	N 810–840°C	570	280	14	180–230	
	Q, T 550–660°C	700–850	430	14	200–255	
070M55	N 810–840°C	600	310	13	200–250	
	Q, T 550–660°C	700–850	415	14	200–255	

Note: N = normalized, Q = quenched, T = tempered. All data refer to large ruling sections.

Mechanical properties of cast irons

Table 3.33 gives the properties in general and 3.34 gives the values of the mechanical properties of specific cast irons. The yield stress of white and grey irons is virtually the same as the tensile strength. For ductile and malleable irons the yield stress is about three-quarters of the tensile strength.

Table 3.33 Mechanical properties of cast irons

Material	Condition	Tensile modulus (GPa)	Tensile strength (MPa)	Elongation (%)	Hardness (BH)	Impact (J)
White	As cast	170	275	0	500	4*
Grey	As cast	100–145	150–400	0.2–0.7	130–300	8–50+
Blackheart	Annealed	170	290–340	6–12	125–140	13–17**
Whiteheart	Annealed	170	270–410	3–10	120–180	2–5**
Pearlitic malleable	Normalized	170	440–570	3–7	140–240	2–10**
Ductile ferritic	As cast	165	370–500	7–17	115–215	5–15**
Ductile pearlitic	As cast	165	600–800	2–3	215–305	2–5**
Abrasion res. alloy	As cast or heat treated	180–200	230–460		400–650	6–12+

Corrosion res.	As cast or annealed	124	90–150		450–520	5–8 +
Heat resistant grey alloy	As cast or annealed	70–110	140–240	1–3	110–740	40–80 +
Heat resistant ductile alloy	As cast or annealed	90–140	370–490	7–40	130–250	4–30 + +

Note: For the impact strengths, • = Charpy unnotched, •• = Charpy notched, + = Izod unnotched, + + = Izod notched.

Table 3.34   Mechanical properties of cast irons

Material	Tensile strength (MPa)	Yield stress (MPa)	Compress. strength (MPa)	Elong- ation (%)	Hard- ness (BH)
Grey irons					
Grade BS 150	150	98	600	0.6	136–167
Grade BS 180	180	117	670	0.5	150–183
Grade BS 220	220	143	770	0.5	167–204
Grade BS 260	260	170	860	0.4	185–226
Grade BS 300	300	195	960	0.3	202–247
Grade BS 350	350	228	1080	0.3	227–278
Grade BS 400	400	260	1200	0.2	251–307
Grade ASTM 20	152		572		156
Grade ASTM 25	179		669		174
Grade ASTM 30	214		752		210
Grade ASTM 35	252		855		212
Grade ASTM 40	293		965		235
Grade ASTM 50	362		1130		262
Grade ASTM 60	431		1293		302
Malleable irons					
Blackheart B30–06	300			6	‹150
Blackheart B32–10	320	190		10	‹150
Blackheart B35–12	350	200		12	‹150
Ferritic ASTM 32510	345	224		10	‹156
Ferritic ASTM 35018	365	241		18	‹156
Whiteheart W35–04	350			4	‹230
Whiteheart W38–12	380	200		12	‹200
Whiteheart W40–05	400	220		5	‹220
Whiteheart W45–07	450	260		7	‹220
Pearlitic P45–06	450	270		6	150–200
Pearlitic P50–06	500	300		5	160–220
Pearlitic P55–04	550	340		4	180–230
Pearlitic P60–03	600	390		3	200–250
Pearlitic P65–02	650	430		2	210–260
Pearlitic P70–02	700	530		2	240–290
Pearlitic ASTM 40010	414	276		10	149–197
Pearlitic ASTM 45008	448	310		8	156–197
Pearlitic ASTM					

45006 Pearlitic ASTM	448	310		6	156–207
50005 Pearlitic ASTM	483	345		5	179–229
60004 Pearlitic ASTM	552	414		4	197–241
70003 Pearlitic ASTM	586	483		3	217–269
80002 Pearlitic ASTM	655	552		2	241–285
90001 Pearlitic ASTM	724	621		1	269–321
Ductile irons					
Grade ASTM 60–40–18	414	276		18	143–187
Grade ASTM 80–60–03	552	414		3	>200
Grade ASTM 60–40–18	414	276		18	167
Grade ASTM 65–45–12	448	310		12	167
Grade ASTM 80–55–06	552	379		6	192
Grade ASTM 100–70–03	689	483		3	
Grade ASTM 120–90–02	827	621		2	331
Abrasion-resistant white alloys					
BS low alloy, 1A, 1B	230–460				400
BS nickel-chromium, 2A–E	250–450				550
BS high chromium, 3A–E	300–450				400–650
Martensitic nickel-chromium					550–650
High chromium white					450–600
Ni hard, 3.5% Ni Cr	280–510				
Corrosion-resistant alloys					
BS Ferritic high silicons	93–154				450–520
High silicon	90–180		690		480–520
High chromium	205–830		690		250–740
High nickel grey	170–310		690–1100		120–250
High nickel ductile	380–480		1240–1380		130–240
Heat-resistant grey alloys					
BS Austenitic					
L-Ni Mn 13 7	140–220				120–150
L-Ni Cu Cr 15 6 2	170–210			2	140–200
L-Ni Cu Cr 15 6 3	190–240			1–2	150–250
L-Ni Cr 20 2	170–210			2–3	120–215
L-Ni Cr 20 3	190–240			1–2	160–250
L-Ni Si Cr 20 5 3	190–280			2–3	140–250

L–Ni Cr 30 3	190–240		1–3	120–215
L–Ni Si Cr 30 5 5	170–240			150–210
L–Ni 35	120–180		1–3	120–140
Medium silicon grey	170–310		620–1040	170–250
High chromium grey	210–620		690	250–500
High nickel grey	170–310		690–1100	130–250
Ni–Cr–Si grey	140–310		480–690	110–210
High aluminium	235–620			180–350
Heat-resistant ductile alloys				
BS Austenitic				
S–Ni Mn 13 7	390–460	210–260	15–25	130–170
S–Ni Cr 20 2	370–470	210–250	7–20	140–200
S–Ni Cr 20 3	390–490	210–260	7–15	150–255
S–Ni Si Cr 20 5 2	370–430	210–260	10–18	180–230
S–Ni 22	370–440	170–250	20–40	130–170
S–Ni Mn 23 4	440–470	210–240	25–45	150–180
S–Ni Cr 30 1	370–440	210–270	13–18	130–190
S–Ni Cr 30 3	370–470	210–260	7–18	140–200
S–Ni Si Cr 30 5 5	390–490	240–310	1–4	170–250
S–Ni 35	370–410	210–240	20–40	130–180
S–Ni Cr 35 3	370–440	210–290	7–10	140–190
Medium silicon ductile	415–690			140–300
High Ni (20) ductile	380–415		1240–1380	140–200

Note: the values quoted for yield stress for grey irons are the 0.1% proof stress, for all other irons for which values are given the 0.2% proof stress.

Mechanical properties of free-cutting steels

Tables 3.35 and 3.36 give the mechanical properties of free-cutting/resulphurized and rephosphorized steels.

Table 3.35 Mechanical properties of AISI resulphurized and rephosphorized steels

Material	Condition	Tensile strength (MPa)	Yield stress (MPa)	Elong-ation (%)	Hard-ness (BH)	Impact Izod (J)
1117	As rolled	490	305	33	140	80
	N 900°C	470	300	34	140	85
	A 855°C	430	280	33	120	94
1118	As rolled	520	320	32	150	110
	N 925°C	480	320	34	140	103
	A 790°C	450	285	35	130	106
1137	As rolled	630	380	28	190	83
	N 900°C	670	400	23	200	64
	A 790°C	585	345	27	170	50
	WQ, T 200°C	1500	1165	5	415	
	WQ, T 650°C	650	530	25	190	
1141	As rolled	675	360	22	190	11
	N 900°C	710	405	23	200	53
	A 790°C	600	350	26	160	34
	Q, T 200°C	1630	1210	6	460	
	Q, T 650°C	710	590	23	220	
1144	As rolled	700	420	21	210	53
	N 900°C	670	400	21	200	43

Q, T 200°C	880	630	17	280
Q, T 650°C	720	500	23	220

Note: Q = oil quenched, WQ = water quenched, T = tempered, N = normalized, A = annealed.

Table 3.36 Mechanical properties of BS free-cutting steels

Material Condition	Tensile strength (MPa)	Yield stress (MPa)	Elong-ation (%)	Hard-ness (BH)	Impact Izod (J)
212M36 Q, T 550–660°C	550–700	340	20	150–210	25
214M15 Q		590		13		35
216M36 Q, T 550–660°C	550–700	380	15	150–210	25
216M44 Q, T 550–660°C	700–850	450	15	200–255	15
220M07 N 900–930°C		360	215	22	>103	
226M44 Q, T 550–660°C	700–850	450	16	200–255	20
230M07 N 900–930°C		360	215	22	>103	

Note: Q = oil quenched, T = tempered, N = normalized.

Mechanical properties of maraging steels

Table 3.37 gives the mechanical properties of maraging steels, see Composition of maraging steels for clarification of the grades.

Table 3.37 Mechanical properties of maraging steels

Grade	Tensile strength (MPa)	0.2% proof (MPa)	Elong-ation (%)	Hard-ness (HV)
200	1390	1340	11	450
250	1700	1620	9	520
300	1930	1810	7	570

Mechanical properties of stainless steels

Tables 3.38 and 3.39 give the mechanical properties of typical stainless and heat resisting steels. The tensile modulus of all the steels is about 200 to 207 GPa or GN m^{-2}.

Table 3.38 Mechanical properties of AISI stainless steels

AISI	Condition	Tensile strength (MPa)	Yield stress (MPa)	Elong-ation (%)
Austenitic				
201	A	790	380	55
301	A	760	280	60
	1/2 H	1450	1310	15
302	A	620	310	50
304	A	580	290	55
	1/2 H	1100	1000	10
309	A	620	310	45
316	A	580	290	50
321	A	620	240	45

Ferritic				
405	A	480	275	20
409	A	470	275	20
430	A	520	345	25
442	A	550	310	20
446	A	550	345	20
Martensitic				
403	A	520	275	30
410	A	520	275	30
420	A	655	345	25
	T 200°C	1760	1380	10
	T 650°C	790	585	20
431	A	860	655	20
	T 200°C	1410	1070	15
	T 650°C	870	655	20
440C	A	760	480	13
	T 300°C	1970	1900	2

Note: A = annealed, 1/2 H = half hard as a result of being cold worked, FH = fully hard, T = tempered. The yield stress is the 0.2% proof stress.

Table 3.39 Mechanical properties of BS stainless steels

BS	Condition	Tensile strength (MPa)	Yield stress (MPa)	Elongation (%)	Hardness (BH)	Impact Izod (J)
Austenitic						
302S31	Soft	510	190	40	183	
303S31	Soft	510	190	40	183	
304S31	Soft	490	195	40	183	
310S31	Soft	510	205	40	207	
316S33	Soft	510	205	40	183	
321S31	Soft	510	200	35	183	
Ferritic						
403S17	AC 740°C	415	280	20	170	
430S15	AC 780°C	430	280	20	170	
Martensitic						
410S21	OQ	1850	1190	3	350	7
	OQ, T 750°C	570	370	33	170	130
416S21	OQ, T 750°C	700	370	15	205	25
420S29	OQ, T 400°C	1500	1360	18	455	16
	OQ, T 700°C	760	630	26	220	95
431S29	OQ, T 650°C	880	695	22	260	34

Note: Soft = softened by heating to about 1000–1100°C, OQ = oil quenched, AC = air cooled, T = tempered.

Oxidation resistance

A factor limiting the use of materials at high temperatures is surface attack, i.e. scaling, due to oxidation. Materials with good oxidation resistance are those that develop an adherent, impervious, oxide layer which can resist the further movement of oxygen. Table 3.40 gives a general idea of the oxidation limits placed on the use of various steels.

Table 3.40 Oxidation limits of steels

Material	Oxidation limit (°C)
Carbon steels	450
0.5% molybdenum steel	500
1% Cr, 0.5% Mo steel	550
12% Cr, Mo, V steel	575
18% Cr, 8% Ni steel	650
19% Cr, 11% Ni, 2% Si steel	900
23% Cr, 20% Ni	1100

Ruling section

The mechanical properties of a steel depend on the size of the cross-section of the material. For this reason, properties are often quoted in terms of the size of the cross-section. The limiting ruling section is the maximum diameter of round bar at the centre of which the specified properties may be obtained. The reason for a difference of mechanical properties occurring for different size bars of the same steel is that during the heat treatment different rates of cooling occur at the centres of the bars due to their differences in size. This results in differences in microstructure and hence differences in mechanical properties. In general, the larger the limiting ruling section the smaller the tensile strength and the greater the percentage elongation. Table 3.41 shows some typical values.

Table 3.41 Effect of ruling section on properties

Material	Ruling section (mm)	Tensile strength (MPa)	Yield stress (MPa)	Elong-ation (%)	Impact Izod (J)
Carbon steel	29	770	590	25	60
	152	700	490	25	40
Ni-Cr-Mo steel	29	1100	930	20	68
	152	1000	850	20	68

Specific gravity

The specific gravity of pure iron at 20°C is 7.88. The addition of alloying elements changes this value by relatively small amounts. For example, carbon, manganese, chromium and aluminium decrease this value, while nickel, molybdenum, cobalt and tungsten increase it. The specific gravity of carbon steels thus tends to be about 7.80, alloy steels about 7.81, ferritic and martensitic stainless steels about 7.7 and austenitic stainless steels about 8.0.

Thermal properties

The thermal properties, i.e. specific heat capacity, thermal conductivity and linear thermal expansivity (or coefficient of linear expansion) vary with temperature. Table 3.42 thus gives average values for temperatures in the region of 20°C.

Table 3.42 Thermal properties of ferrous alloys

Material	Specific ht. cap. (J kg^{-1} °C^{-1})	Thermal conduct. (W m^{-1} °C^{-1})	Thermal expans. (10^{-6} °C^{-1})
Alloy steel	510	37	12
Carbon steel	480	47	15
Cast iron, grey	265–460	53–44	11

Stainless, ferritic	510	26	11
Stainless, martensitic	510	25	11
Stainless, austenitic	510	16	16

Note: To convert J kg⁻¹ °C⁻¹ to cal g⁻¹ °C⁻¹ multiply by 2.39 × 10⁻⁴. To convert W m⁻¹ °C⁻¹ to cal cm⁻¹ °C⁻¹ s⁻¹ multiply by 2.39 × 10⁻³.

Tool steel properties

Tool steels can be compared on the basis of their hardening and behaviour in use characteristics. Hardening characteristics are the depth of hardening (as a measure of hardenability), the risk of cracking during the hardening operation, the amount of distortion during the hardening operation, and resistance to decarburization (when the steel is heated to hardening temperatures there can be a loss of carbon from surface layers, such a loss leading to a softer surface). Behaviour in use characteristics are resistance to heat softening at temperatures attained during use of the tool, wear resistance, toughness and machinability. Table 3.43 shows a comparison of these properties for tool steels. Table 3.44 gives values of cold and hot hardness for tool steels typical of each type of material. In general, D and O tool steels are subject to less distortion than other grades, though distortion can be kept low with O grade. S grade steels have toughness under impact conditions. H steels are designed for use at elevated temperatures. The high speed steels have red hardness, i.e. they retain their hardness and hence cutting edge when cutting at speed and hence becoming hot.

See Coding system for tool steels, Composition of tool steels and Uses of tool steels.

Table 3.43 Comparison of tool steel properties

AISI	Quench medium	Depth of hard.	Risk of crack.	Dist- ortion	Resis. to decarb.	Resis. to soft.	Wear resis.	Tough- ness	Machin- ability
Water hardening									
W1	W	P	H	H	VH	L	F	H	VG
W2	W	P	H	H	VH	L	F	H	VG
W5	W	P	H	H	VH	L	F	H	VG
Shock-resisting									
S1	O	M	L	M	M	M	F	VG	F
S2	W	M	H	H	L	L	F	VG	F
S5	O	M	L	M	L	L	F	VG	F
S7	A	G	VL	L	M	H	F	VG	F
Oil-hardening									
O1	O	M	L	L	H	L	M	M	G
O2	O	M	L	L	H	L	M	M	G
O6	O	M	L	L	H	L	M	M	VG
O7	O	M	L	L	H	L	M	M	G
Air-hardening									
A2	A	G	VL	VL	M	H	G	M	F
A3	A	G	VL	VL	M	H	VG	M	F
A4	A	G	VL	VL	M/H	M	G	M	P/F
A6	A	G	VL	VL	M/H	M	G	M	P/F
A7	A	G	VL	VL	M	H	VG	L	P
A8	A	G	VL	VL	M	H	F/G	H	F
A9	A	G	VL	VL	M	H	F/G	H	F
A10	A	G	VL	VL	M/H	M	G	M	F/G

High carbon-high chromium steels

D2	A	G	VL	VL	M	H	VG	L	P
D3	O	G	L	VL	M	H	VG	L	P
D4	A	G	VL	VL	M	H	VG	L	P
D5	A	G	VL	VL	M	H	VG	L	P
D7	A	G	VL	VL	M	H	VG	L	P

Hot-work chromium steels

H10	A	G	VL	VL	M	H	F	G	F/G
H11	A	G	VL	VL	M	H	F	VG	F/G
H12	A	G	VL	VL	M	H	F	VG	F/G
H13	A	G	VL	VL	M	H	F	VG	F/G
H14	A	G	VL	L	M	H	F	G	G
H19	A	G	L	L	M	H	F/G	G	G

Hot-work tungsten steels

H21	A/O	G	L	L	M	H	F/G	G	G
H22	A/O	G	L	L	M	H	F/G	G	G
H23	A/O	G	L	L	M	VH	F/G	F	G
H24	A/O	G	L	L	M	VH	G	F	G
H25	A/O	G	L	L	M	VH	F	G	G
H26	A/O	G	L	L	M	VH	G	F	G

Hot-work molybdenum steels

H42	A/O	G	L	L	M	VH	G	F	G

High-speed tungsten steels

T1	O	G	L	M	H	VH	VG	L	F
T1	A/S	G	L	L	H	VH	VG	L	F
T2	O	G	L	M	H	VH	VG	L	F
T2	A/S	G	L	L	H	VH	VG	L	F
T4	O	G	M	M	M	VH	VG	L	F
T4	A/S	G	M	L	M	VH	VG	L	F
T5	O	G	M	M	L	VH	VG	L	F
T5	A/S	G	M	L	L	VH	VG	L	F
T6	O	G	M	M	L	VH	VG	L	P/F
T6	A/S	G	M	L	L	VH	VG	L	P/F

High-speed molybdenum steels

M1	O	G	M/H	M	L	VH	VG	L	F
M1	A/S	G	M/H	L	L	VH	VG	L	F
M2	O	G	M/H	M	M	VH	VG	L	F
M2	A/S	G	M/H	L	M	VH	VG	L	F
M4	O	G	M/H	M	M	VH	VG	L	P/F
M4	A/S	G	M/H	L	M	VH	VG	L	P/F
M6	O	G	M/H	M	L	VH	VG	L	F
M6	A/S	G	M/H	L	L	VH	VG	L	F
M7	O	G	M/H	M	L	VH	VG	L	F
M7	A/S	G	M/H	L	L	VH	VG	L	F
M10	O	G	M/H	M	L	VH	VG	L	F
M10	A/S	G	M/H	L	L	VH	VG	L	F
M30	O	G	M/H	M	L	VH	VG	L	F
M30	A/S	G	M/H	L	L	VH	VG	L	F
M33	O	G	M/H	M	L	VH	VG	L	F
M33	A/S	G	M/H	L	L	VH	VG	L	F
M34	O	G	M/H	M	L	VH	VG	L	F
M34	A/S	G	M/H	L	L	VH	VG	L	F
M36	O	G	M/H	M	L	VH	VG	L	F
M36	A/S	G	M/H	L	L	VH	VG	L	F
M41	O	G	M/H	M	L	VH	VG	L	F
M41	A/S	G	M/H	L	L	VH	VG	L	F
M42	O	G	M/H	M	L	VH	VG	L	F
M42	A/S	G	M/H	L	L	VH	VG	L	F
M43	O	G	M/H	M	L	VH	VG	L	F
M43	A/S	G	M/H	L	L	VH	VG	L	F

M44	O	G	M/H	M	L	VH	VG	L	F
M44	A/S	G	M/H	L	L	VH	VG	L	F
M46	O	G	M/H	M	L	VH	VG	L	F
M46	A/S	G	M/H	L	L	VH	VG	L	F
M47	O	G	M/H	M	L	VH	VG	L	F
M47	A/S	G	M/H	M	L	VH	VG	L	F

Note: Quench media; W = water, O = oil, A = air, S = salt.
Depth of hardness; G = good depth, M = medium depth, P = poor depth.
Risk of cracking; H = high risk, M = medium risk, L = low risk.
Distortion during hardening; H = high amount, M = medium amount, L = low amount, VL = very low amount.
Resistance to decarburization; VH = very high resistance, H = high resistance, M = medium resistance, L = low or little resistance.
Resistance to the softening effects of heat; VH = very high resistance, H = high resistance, M = medium resistance, L = low resistance.
Wear resistance; VG = very good, G = good, M = medium, F = fair, L = low or little.
Toughness; VH = very high, H = high, M = medium, L = low.
Machinability; VG = very good, G = good, F = fair, P = poor.

Table 3.44 Hardness values of tool steels

Type of tool steel	AISI	Hardness R_c	
		at 20°C	at 560°C
Water-hardening	W1	63	10
Shock-resisting	S1	60	20
Oil-hardening	O1	63	20
Air-hardening	A2	63	30
High C–High Cr	D2	62	35
High speed tungsten	T1	66	52
High speed molybdenum	M10	65	52

3.5 Uses

Uses of alloy steels

Tables 3.45 and 3.46 give some typical uses of a range of the more commonly used alloy steels.

Table 3.45 Uses of BS alloy steels

Material	Typical uses

Manganese steels

150M19 Makes good welds and used where tensile strengths up to about 540 MPa are required, e.g. large diameter shafts, lifting gear.

150M36 Less weldable than 150M19 but tensile strength up to about 1000 Pa in small sections. Used for axles, levers, gun parts.

Chromium steels

530M40 A commonly used alloy steel where tensile strengths less than 700 MPa are required, e.g. crankshafts, axles, connecting rods.

Manganese-molybdenum

605M36 Widely used where tensile strengths up to 850 MPa are required, e.g. heavy duty shafts, high tensile studs and bolts, connecting rods.

608M38   Tensile strength requirements up to about 1000 MPa,
e.g. differential shafts, connecting rods, high-tensile
studs and bolts.

Nickel–chromium

653M31   Components requiring tensile strengths up to about
900 MPa, e.g. differential shafts, high-tensile studs
and bolts, connecting rods.

Chromium–molybdenum

708M40   Widely used where tensile strengths up to about 850
MPa in large sections and 930 MPa in small sections
are required.

709M40   Has similar applications to 605M36 but in larger
sections. Has good wear resistance. Used for high
tensile bolts, track pins.

722M24   Can be nitrided to give a surface hardness of 900 HV.
Tough. Used for heavy duty crankshafts, machine tool
parts.

Nickel–chromium–molybdenum

817M40   Used where tensile strengths up to 1000 MPa are
required in large sections and 1500 MPa in small
sections and where good shock and fatigue properties
are necessary, e.g. gears.

826M40   Can be used for components requiring tensile
strengths up to 1400 MPa.

835M30   Used where tensile strengths up to 1500 MPa are
required.

Manganese–nickel–chromium–molybdenum

945M38   An alternative to 817M40.

Table 3.46 Uses of AISI alloy steels

Material	Uses
Manganese steels	
1330	Used where higher strengths than carbon steels are required, e.g. axles, shafts, tie rods.
1340	Similar uses to 1330.
Molybdenum steels	
4037	Used for rear axle drive pinions and gears.
Chromium–molybdenum	
4130	Has increased hardenability, strength and wear resistance when compared with comparable carbon steel. Can be oil-quenched instead of water-quenched. Used for aircraft structural parts, car axles, pressure vessels.
4140	Similar uses to 4130.
Nickel–molybdenum steels	
4620	Used for transmission gears, shafts, roller bearings.
4820	Similar uses to 4620.
Chromium steels	
5130	Carburizing steel.
5140	Has increased hardenability, strength and wear resistance when compared with comparable carbon steel.
5160	Used as a spring steel.
Nickel–chromium–molybdenum	
4340	Has higher elastic limit, impact strength, fatigue resistance and hardenability than comparable carbon steel. Used for heavy duty, high strength parts, e.g. landing gears.

8620	Carburizing steel.
8630	Used for shafts and forgings requiring high strengths.
8650	Similar uses to 8630.

Silicon steels

9260	Used for leaf springs.

Uses of carbon steels

Low carbon sheet steel, with about 0.06 to 0.12% carbon, is widely used for car body work and tin plate. It is easily formed and welded and is cheap. It has sufficient strength for such applications, high strength not being a requirement. Such steels are non-heat-treatable and the final condition is as a result of temper-rolling.

Hardenable carbon steels can be considered to fall into three groups: low carbon steels with 0.10 to 0.25% carbon; medium carbon steels with 0.25 to 0.55% carbon and high carbon steels with 0.55 to 1.00% carbon. Low carbon steels are not usually quenched and tempered but may be carburized or case hardened. Such steels are used for engine fans, pulley wheels, and other lightly stressed parts. Medium carbon steels offer a range of properties after quenching and tempering. They have a wide range of uses, e.g. as shafts, or parts in car transmissions, suspensions and steering. High carbon steels are more restricted in their uses since they have poor formability and weldability, as well as being more costly. They are used in the quenched and tempered state and have applications such as clutch cams.

Uses of cast irons

Table 3.47 outlines some of the uses made of cast irons. Grey irons have very good machinability and stability, good wear resistance but poor tensile strength and ductility. They have good strength in compression but poor impact properties. White iron has excellent abrasion resistance and is very hard. It is however virtually unmachinable and so has to be cast to its final shape. Malleable irons have good machinability and stability with higher tensile strengths and ductility than grey irons, also better impact properties. Ductile irons have high tensile strength with reasonable ductility. Their machinability and wear characteristics are good but not as good as those of grey irons.

Table 3.47 Uses of cast irons

Material	Uses
Grey irons	Water pipes, motor cylinders and pistons, machine castings, crankcases, machine tool beds, manhole covers
White iron	Wear-resistant parts, such as grinding mill parts and crusher equipment
Malleable irons	Wheel hubs, pedals, levers, general hardware, bicycle and motor-cycle frame fittings
Ductile irons	Heavy duty piping, crankshafts
Abrasion resistant white	Abrasive materials handling equipment, ore crushing jaws
Corrosion resistant alloy	Uses requiring resistance to oxidizing acids. High silicon content results in very brittle material which is difficult to machine
Heat resistant grey alloy	High resistance to heat and some resistance to corrosion. Ni Mn 13 7 is non-magnetic.

Heat resistant ductile alloy	Tough and ductile to low temperatures. The high ductile alloy nickel alloys have low thermal expansivities.

Uses of stainless steels

Table 3.48 gives some typical uses of stainless steels. In general, austenitic steels are widely used because of their excellent corrosion resistance and formability, finding uses at both normal and high temperatures. Ferritic steels are mainly used as general construction materials where their good corrosion and heat-resistant properties can be exploited. They are however less used than austenitic steels since they have poor weldability, lack ductility and are notch sensitive. Martensitic steels have poorer corrosion resistance than austenitic or ferritic steels. Their tensile properties are controlled by the heat treatment they undergo.

Table 3.48 Uses of stainless steels

AISI	BS	Uses
Austenitic		
201		A low nickel equivalent of 301, used for car wheel covers and trim.
301		Used where high strength with high ductility required, e.g. car wheel covers and trim, railway wagons, fasteners.
302	302S31	Widely used for fabrications, domestic and decorative purposes, e.g. hospital and household appliances, food handling equipment, springs, tanks, signs.
304	304S31	Used for chemical and food processing equipment, gutters, downspouts and flashings.
309		Has high temperature strength and oxidation resistance, hence uses in heat treatment equipment, heat exchangers, oven liners.
	310S31	Used for high temperature applications such as furnace components and superheater tubes.
316	316S33	High corrosion and creep resistance. Used for chemical handling equipment.
321	321S31	The titanium in this alloy stabilizes it, hence its use where weldments are subject to severely corrosive conditions and/or high temperatures, e.g. aircraft exhaust manifolds, boiler shells, pressure vessels.
Ferritic		
	403S17	A soft and ductile steel. Used for domestic utensils, pressings, drawn components, spun components.
405		Used for structures requiring welding. Uses include quenching racks.
409		A general purpose steel. Used for car exhaust systems, tanks for agricultural sprays or fertilizers.
430	430S17	A general purpose steel. Used for decorative trims, dishwashers, heaters, acid tanks.
442		Used for high temperature applications, e.g. furnace parts.
446		Used for high temperature applications, e.g. combustion chambers, glass moulds, heaters, valves.

Martensitic

403	420S29	Used for steam turbine blades.
410	410S21	A general purpose steel. Used for machine parts, cutlery, screws, bolts, valves.
	416S21	Similar uses to 410.
420	420S37	Used for springs, machine parts, scissors, bolts.
431	431S29	Used where very high mechanical properties are needed, e.g. aircraft fittings.
440C		Has the highest hardness of hardenable stainless steels. Hence used for bearings, races, nozzles, valve parts.

Uses of tool steels

Table 3.49 gives uses of some of the more commonly used tool steels.

Table 3.49 Uses of tool steels

AISI	Uses
Water-hardening	
W1 (0.9 C)	Gauges, chisels, punches, shear blades, rivet sets, forging dies, blanking tools.
(1.0 C)	Large drills, cutters, reamers, shear blades, woodworking tools, cold heading tools, punches, countersinks, blanking dies.
(1.2 C)	Twist drills, cutters, reamers, woodworking tools, taps, files, lathe tools.
W2	Large drills, cutters, reamers, woodworking tools, shear blades, cold heading tools.
W5	Heavy stamping and draw dies, reamers, large punches, razor blades, cold forming rolls and dies.
Shock-resisting	
S1	Concrete drills, pneumatic tools, shear blades, bolt header dies, mandrels, pipe cutters, forging dies.
S2	Hand and pneumatic chisels, shear blades, forming tools, mandrels, stamps, spindles, screw driver bits, tool shanks.
S5	Hand and pneumatic chisels, shear blades, forming tools, mandrels, stamps, spindles, punches, rotary shears, pipe cutters.
S7	Shear blades, punches, chisels, forming dies, blanking dies, chuck jaws, pipe cutters.
Oil-hardening	
O1	Blanking, drawing and trimming dies, plastic moulds, shear blades, taps, reamers, gauges, bushes, punches, jigs, paper knives.
O2	Blanking, stamping, trimming and forming dies and punches, taps, reamers, gauges, bushes, jigs, circular cutters and saws.
O6	Blanking and forming dies, blanking and forming punches, mandrels, cold forming rollers, gauges, taps, tool shanks, jigs.
O7	Blanking and forming dies, mandrels, gauges, taps, drills, paper knives, woodworking tools.
Air hardening	
A2	Thread rolling dies, extrusion dies, trimming, blanking and coining dies, mandrels, shear blades, spinning and forming rolls, gauges, burnishing tools, plastic moulds, bushes, punches.

A3	As A2.
A4	Blanking, forming and trimming dies, punches, shear blades, mandrels, forming rolls, gauges, punches.
A6	Blanking, forming, coining and trimming dies, punches, shear blades, mandrels, plastic moulds.
A7	Liners for shot blasting equipment, forming dies, gauges, drawing dies.
A8	Blanking, coining and forming dies, shear blades.
A9	Cold heading dies, coining dies, forming dies and rolls, hot-working tools such as punches, mandrels, extrusion tools, hammers.
A10	Blanking and forming dies, gauges, punches, forming rolls, wear plates.

High carbon–high chromium steels

D2	Blanking, drawing and cold-forming dies, thread rolling dies, shear blades, burnishing tools, punches, gauges, broaches, mandrels, cutlery.
D3	Blanking, drawing and cold-forming dies, thread rolling dies, shear blades, burnishing tools, punches, gauges, crimping dies.
D4	Blanking dies, thread rolling dies, wire drawing dies, forming tools and rolls, punches, dies for deep drawing.
D5	Blanking, coining and trimming dies, shear blades, cold forming dies, shear blades, punches, quality cutlery.
D7	Wire drawing dies, deep drawing dies, ceramic tools and dies, sand-blasting liners.

Hot-work chromium steels

H10	Hot extrusion and forging dies, mandrels, punches, hot shears, die holders and inserts.
H11	Die-casting dies for light alloys, forging dies, punches, piercing tools, sleeves, mandrels.
H12	Extrusion dies, punches, mandrels, forging die inserts.
H13	Die-casting dies and inserts, extrusion dies, forging dies and inserts.
H14	Aluminium and brass extrusion dies, forging dies and inserts, hot punches.
H19	Extrusion dies and inserts, forging dies and inserts, mandrels, hot punch tools.

Hot-work tungsten steels

H21	Mandrels, hot blanking dies, hot punches, extrusion dies and die-casting dies for brass, gripper dies, hot headers.
H22	Mandrels, hot blanking dies, hot punches, extrusion dies, gripper dies.
H23	Extrusion and die-casting dies for brass.
H24	Hot blanking and drawing dies, trimming dies, hot press dies, hot forming dies, hot heading dies, extrusion dies.
H25	Hot forming dies, die-casting and forging dies, shear blades, gripper dies, mandrels, punches, hot swaging dies.
H26	Hot blanking dies, hot punches, hot trimming dies, gripper dies, extrusion dies for brass and copper.

Hot-work molybdenum steels

H82	Cold trimming dies, cold header and extrusion dies and die inserts, hot upsetting dies, hot punches, mandrels.

High-speed tungsten steels

T1	Drills, taps, reamers, hobs, broaches, lathe and planer tools, punches, burnishing dies, milling cutters.
T2	Lathe and planer tools, milling cutters, broaches, reamers.
T4	Lathe and planer tools, boring tools, drills, milling cutters, broaches.
T5	Lathe and planer tools, heavy duty tools requiring red hardness.
T6	Heavy duty lathe and planer tools, drills, milling cutters.

High-speed molybdenum steels

M1	Drills, taps, reamers, milling cutters, hobs, punches, lathe and planer tools, form tools, broaches, saws, routers, woodworking tools.
M2	Drills, taps, reamers, milling cutters, hobs, lathe and planer tools, form cutters, broaches, saws, cold forming tools.
M4	Heavy duty broaches, reamers, milling cutters, lathe and planer tools, form cutters, blanking dies and punches, swaging dies.
M6	Lathe and planer tools, form tools, milling cutters, boring tools.
M7	Drills, taps, reamers, routers, saws, milling cutters, lathe and planer tools, woodworking tools, punches, hobs.
M10	As M7.
M30	Lathe tools, milling cutters, form tools.
M33	Drills, taps, lathe tools, milling cutters.
M34	As M33.
M36	Heavy duty lathe and planer tools, milling cutters, drills.
M41	Drills, milling cutters, lathe tools, hobs, broaches, form cutters.
M42	As M41.
M43	As M41.
M44	As M41.
M46	As M41.
M47	As M41.

4 Aluminium Alloys

4.1 Materials

Aluminium

Pure aluminium is a weak, very ductile, material. The mechanical properties depend not only on the purity of the aluminium but also upon the amount of work to which it has been subject. A range of tempers is thus produced by different amounts of work hardening. It has an electrical conductivity about two-thirds that of copper but weight for weight is a better conductor. Aluminium has a great affinity for oxygen and any fresh metal in air rapidly oxidizes to give a thin layer of the oxide on the metal surface. This layer is not penetrated by oxygen and so protects the metal from further attack.

Aluminium alloys

Aluminium alloys can be divided into two groups, wrought alloys and cast alloys. Each of these can be divided into two further groups: those alloys which are not heat treatable and those which can be heat treated. The non-heat-treatable alloys have their properties controlled by the extent of the working to which they are subject. A range of tempers is thus produced. The heat-treatable alloys have their properties controlled by heat treatment. Like aluminium, the alloys have a low density, good electrical and thermal conductivity and a high corrosion resistance. The corrosion resistance properties of sheet alloy are improved by cladding it with layers of unalloyed aluminium.

The main alloying elements used with aluminium are copper, iron, manganese, magnesium, silicon and zinc. Table 4.1 shows the main effects of such elements.

Table 4.1 Alloying elements used with aluminium

Element	Main effects
Copper	Up to about 12% increases strength. Precipitation hardening possible. Improves machinability.
Iron	Small percentages increase strength and hardness and reduce chances of hot cracking in castings.
Manganese	Improves ductility. Improves, in combination with iron, the castability.
Magnesium	Improves strength. Precipitation hardening becomes possible with more than 6%. Improves corrosion resistance.
Silicon	Improves castability, giving an excellent casting alloy. Improves corrosion resistance.
Zinc	Reduces castability. Improves strength when combined with other alloying elements.

Cast alloys

An alloy for use in the casting process must flow readily to all parts of the mould and on solidifying it should not shrink too much and any shrinking should not result in fractures. The choice of alloy is affected by the casting process used. In sand casting the cooling rate is relatively slow, while with die casting when the metal is injected under pressure it is much faster. The cooling rate affects the strength of the finished casting. Thus an aluminium alloy that might be suitable for sand casting may not be the most appropriate one for die casting.

A family of alloys that is used in the 'as cast' condition, i.e. with no heat treatment, has silicon as the major alloying element (see Figure 4.1 for the equilibrium diagram). The addition of the silicon improves fluidity. The eutectic for aluminium–silicon alloys has a composition of 11.6% silicon. An alloy of this composition changes from the liquid to solid state without any change in temperature, and alloys close to this composition solidify over a small temperature range. This makes them particularly suitable for die casting where a quick change from liquid to solid is required in order that a rapid ejection from the die can permit high output rates. The microstructure of the eutectic composition is rather coarse and rather poor mechanical properties result for the casting. The structure can however be modified and made finer by the addition of about 0.005 to 0.15% sodium. This also causes the eutectic composition to change to about 14% silicon (see the dashed line in Figure 4.1).

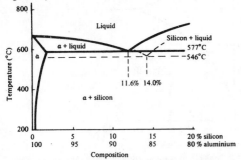

Figure 4.1 The aluminium–silicon equilibrium diagram

Other cast alloys that are not heat treated are aluminium–silicon–copper and aluminium–magnesium–manganese alloys. The aluminium–silicon–copper alloys can be both sand cast and die cast, the aluminium–magnesium–manganese alloys are however only suitable for sand casting.

The addition of copper to the aluminium–silicon alloys allows the casting to be heat treated. The addition of small amounts of magnesium to aluminium–silicon alloys also gives a heat-treatable material.

See Coding system for composition of cast alloys, Coding system for temper, Composition of cast alloys, Annealing, Heat treatment of cast alloys, Electrical properties, Mechanical properties, Uses of cast alloys.

Wrought alloys

Commonly used non-heat-treatable wrought aluminium alloys are aluminium–manganese and aluminium–magnesium (Figure 4.2 shows the relevant part of the equilibrium diagram). A widely used group of heat-treatable alloys is based on aluminium–copper (Figure 4.3 shows the relevant portion of the equilibrium diagram). When such an alloy, say 3% copper, is slowly cooled, the structure at 540°C of a solid solution of the alpha phase, gives a precipitate of a copper–aluminium compound when the temperature falls below the solvus. The result at room temperature is an alpha solid solution with this precipitate. The precipitate is rather coarse, but this can be changed by heating to about 500°C, soaking at that temperature, and then quenching to give a supersaturated solid solution, just alpha phase with no precipitate. This treatment is known as solution treatment and is an unstable situation. With time a fine precipitate is produced. Heating to about 165°C for 10 hours hastens this process. This is called aging and the entire process precipitation hardening, since the result is a stronger, harder material.

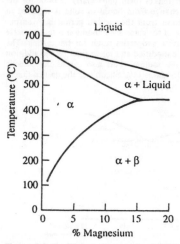

Figure 4.2 Equilibrium diagram for aluminium–magnesium alloys

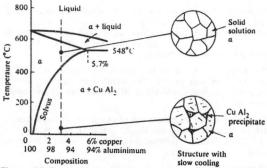

Figure 4.3 Figure 4.3 Equilibrium diagram for aluminium–copper alloys

Other heat-treatable wrought aluminium alloys are aluminium–copper–magnesium–silicon alloys with precipitates forming of an aluminium–copper compound and an aluminium–copper–magnesium compound. Other alloys are based on aluminium–magnesium–silicon and aluminium–zinc–magnesium–copper.

See Coding system for composition of wrought alloys, Coding system for temper, Annealing, Heat treatment of wrought alloys, Electrical properties, Fabrication properties, Mechanical properties, Thermal properties, Forms of material, Uses of wrought alloys.

4.2 Codes and compositions

Coding system for composition of cast alloys

The Aluminium Association system used for the coding of cast aluminium alloys uses four digits, with the last digit separated from the other three by a decimal point. The first digit represents the principal alloying element (see Table 4.2). The second and third digits identify specific alloys within the group. The fourth digit indicates the product form, a 0 indicating a casting and a 1 an ingot. A modification of the original composition is indicated by a letter before the numerical code.

Table 4.2 Coding system for cast alloy composition: Aluminium Association

Digits	Principal alloying element(s)
1XX.X	Aluminium of 99.00% minimum purity
2XX.X	Copper
3XX.X	Silicon plus copper or magnesium
4XX.X	Silicon
5XX.X	Magnesium
6XX.X	Unused series of digits
7XX.X	Zinc
8XX.X	Tin
9XX.X	Other element

In the British Standards (BS) cast alloys are specified according to BS 1490 and the code used consists of the letters LM followed by a number. The number is used to indicate a specific alloy. Table 4.3 indicates the relationship between BS cast alloys and those of the Aluminium Association (AA) for those alloys commonly used and referred to later in this chapter.

Table 4.3 Equivalence of BS and AA numbers for cast alloys

BS	AA
LM4	319.0
LM5	514.0
LM9	A360.0
LM10	520.0
LM12	222.0
LM13	336.0
LM16	355.0
LM18	443.0
LM20	413.0
LM21	319.0
LM24	A380.0
LM25	356.0
LM30	390.0

Coding system for composition of wrought alloys

The coding system used for composition of wrought aluminium alloys is that of the Aluminium Association. The system uses four digits with the first digit representing the principal alloying element (see Table 4.4). The second digit is used to indicate modifications to impurity limits. The last two digits, for the 1XXX alloys, indicate the aluminium content above 99.00% in hundredths. For alloys in other series of digits, the last two digits are used purely to iden specific alloys.

Table 4.4 Coding system for wrought alloy composition

Digits	Principal alloying element(s)
1XXX	Aluminium of 99.00% minimum purity
2XXX	Copper
3XXX	Manganese
4XXX	Silicon
5XXX	Magnesium
6XXX	Magnesium and silicon
7XXX	Zinc
8XXX	Other elements
9XXX	Unused series of digits

Coding system for temper

The complete specification of an alloy, cast or wrought, requires a specification of the temper, i.e. degree of cold work or heat treatment, in conjunction with the composition specification. Table 4.5 shows the basic designations used according to the American system. The British system is a variation of this and is shown in Table 4.7. The basic letter designations can have numbers after them to indicate degrees of cold working or heat treatment, Table 4.6 showing the more commonly used ones. The full specification thus consists of the alloy composition codes followed by a dash and then the temper designation, e.g. 1060–H14.

Table 4.5 Basic American temper codes

Code	Temper
F	As fabricated.
O	Annealed.
H	Strain hardened, wrought products only.
T	Heat treated to produce stable tempers other than those specified by O or H.
W	Solution heat treated. Used only with those alloys that naturally age harden.

Table 4.6 Subdivisions of temper codes

Subdivision	Temper
Strain hardened	
H1	Strain hardened only, no heat treatment. The degree of hardening is indicated by a second digit 1 to 8. 1 indicates the least amount of hardening, 8 the most. H12 is quarter hard, H14 half hard, H16 three-quarters hard, H18 full hard.
H2	Strain hardened and partially annealed. The degree of hardening, after the annealing, is indicated by a second digit 1 to 8, as with H1
H3	Strain hardened and stabilized by a low temperature heat treatment. The degree of hardening, before the stabilization, is indicated by a second digit 1 to 8, as with H1.
Heat treated	
T1	Cooled from an elevated temperature shaping process and naturally aged.
T2	Cooled from an elevated temperature shaping

	process, cold worked, and naturally aged
T3	Solution heat treated, cold worked and naturally aged.
T4	Solution heat treated and naturally aged. T42 indicates the material is solution treated from the O or F temper.
T5	Cooled from an elevated temperature shaping process and then artificially aged.
T51	Stress relieved by stretching, after solution treatment or cooling from an elevated temperature shaping. T510 are products that receive no further straightening after stretching, T511 is where some minor straightening may occur.
T52	Stress relieved by compressing, after solution treatment or cooling from an elevated temperature shaping.
T54	Stress relieved by combined stretching and compressing.
T6	Solution heat treated and then artificially aged. T62 indicates the material is solution heat treated from the O or F temper.
T7	Solution heat treated and stabilized.
T8	Solution heat treated, cold worked, and artificially aged.
T9	Solution heat treated, artificially aged and then cold worked.
T10	Cooled from an elevated temperature shaping process, cold worked and then artificially aged.

Table 4.7 British (BS) temper codes

BS Code	American equivalent	Temper
M	F	As manufactured.
O	O	Annealed.
H	H	Strain hardened. Followed by a number 1 to 8 to indicate degree of hardening, as with the American codes.
TB	T4	Solution treated and naturally aged. No cold work after heat treatment, other than possibly some slight flattening or straightening.
TB7	T7	Solution treated and stabilized.
TD	T3	Solution heat treated, cold worked and naturally aged.
TE	T10	Cooled from an elevated temperature shaping process and precipitation hardened.
TF	T6	Solution heat treated and precipitation hardened.
TF7		Solution treated, precipitation hardened and stabilized.
TH	T8	Solution heat treated, cold worked and then precipitation hardened.
TS	T	Thermally treated to improve dimensional stability.

Composition of cast alloys

Table 4.8 shows the compositions of commonly used Aluminium Association aluminium cast alloys, and Table 4.9 those for British Standards.

Table 4.8 Composition of AA cast alloys

AA no.	Process	Composition percentages				
		Si	Cu	Mg	Mn	Other
Non-heat-treatable						
Aluminium–copper alloys						
208.0	S*	3.0	4.0			
213.0	S, P	2.0	7.0			
Aluminium–silicon–copper/magnesium alloys						
308.0	S, P	5.5	4.5			
319.0	S, P*	6.0	3.5			
360.0	D	9.5		0.50		<2.0 Fe
A360.0	D	9.5		0.50		<1.3 Fe
380.0	D	8.5	3.5			<2.0 Fe
Aluminium–silicon alloys						
413.0	D	12.0				<2.0 Fe
C443.0	D	5.2	<0.6			<2.0 Fe
Heat-treatable						
Aluminium–copper alloys						
222.0	S, P		10.0	0.25		
242.0	S, P		4.0	1.5		2.0 Ni
295.0	P	0.8	4.5			
Aluminium–silicon–magnesium alloys						
355.0	S, P	5.0	1.2	<0.50		<0.6 Fe, <0.35 Zn
356.0	S, P	7.0	<0.25	0.32	<0.35	
357.0	S, P	7.0		0.50		

Note: Process S = sand casting, P = permanent mould casting, D = die casting. With S and P the metal is poured into the mould under gravity, with D it is under pressure. D gives the fastest cooling rate, S the slowest. * = heat treatment is optional.

Table 4.9 Composition of BS cast alloys

BS	Process	Mean composition percentages				
		Si	Cu	Mg	Mn	Other
LM4	S, C	5.0	3.0	<0.15	0.4	<1.0 Fe, <0.3 Ni, <0.5 Zn
LM5	S, C	<0.3	<0.1	4.5	0.5	<0.6 Fe, <0.3 Ni, <0.1 Zn
LM6	S, C	11.5	<0.1	<0.1	<0.5	<0.6 Fe, <0.1 Ni, <0.1 Zn
LM9	S, C	11.5	<0.1	0.4	0.5	<0.6 Fe, <0.1 Ni, <0.1 Zn
LM10	S, C	<0.25	<0.1	10.2	<0.1	<0.35 Fe, <0.1 Ni, <0.1 Zn
LM12	C	<2.5	10.0	0.3	<0.6	<1.0 Fe, <0.5 Ni, <0.8 Zn
LM13	S, C	11.0	0.9	1.2	<0.5	<1.0 Fe, <1.5 Ni, <0.5 Zn
LM16	S, C	5.0	1.3	0.5	<0.5	<0.6 Fe, <0.1 Zn, <0.25 Ni
LM18	S, C	5.3	0.1	<0.1	0.5	0.6 Fe, <0.1 Ni, <0.1 Zn
LM20	C	11.5	<0.4	<0.2	<0.5	<1.0 Fe, <0.1 Ni, <0.2 Zn
LM21	S, C	6.0	4.5	0.2	0.4	<1.0 Fe, <0.3 Ni, <2.0 Zn
LM24	C	8.0	3.5	<0.3	<0.5	<1.3 Fe, <0.5 Ni, <0.1 Zn
LM25	S, C	7.0	<0.1	0.4	<0.3	<0.5 Fe, <0.1 Ni, <0.1 Zn
LM30	C	17.0	4.5	0.6	<0.3	<1.1 Fe, <0.1 Ni, <0.2 Zn

Note: All the alloys also contain about 0.1–0.3 Pb, 0.05–0.10 Sn and 0.2 Ti. S = sand casting, C = chill casting, i.e. where the die causes a more rapid cooling than sand casting.

Composition of wrought alloys

Table 4.10 shows the composition of commonly used wrought alloys specified to the Aluminium Association codes.

Table 4.10 Composition of wrought alloys

AA no.	Composition percentages						
	Al	Mn	Mg	Cu	Si	Cr	Other
Non-heat-treatable							
Unalloyed aluminium							
1050	⟩99.50						Cu, Si, Fe
1060	⟩99.60						Cu, Si, Fe
1100	⟩99.00			0.12			Si, Fe
1200	⟩99.00						Cu, Si, Fe
Aluminium–manganese alloys							
3003	98.6	1.2		0.12			
3004	97.8	1.2	1.0				
3105	99.0	0.55	0.50				
Aluminium–magnesium alloys							
5005	99.2		0.8				
5050	98.6		1.4				
5052	97.2		2.5			0.25	
5083	94.7	0.7	4.4			0.15	
5086	95.4	0.4	4.0			0.15	
5154	96.2		3.5			0.25	
5252	97.5		2.5				
5454	96.3	0.8	2.7			0.12	
5456	93.9	0.8	5.1			0.12	
Heat-treatable							
Aluminium–copper alloys							
2011	93.7			5.5			0.4 Bi, 0.4 Pb
2014	93.5	0.8	0.5	4.4	0.8		
2024	93.5	0.6	1.5	4.4			
2219	93.0	0.3		6.3			0.06 Ti, 0.1 V, 0.18 Zr
2618	93.7		1.6	2.3	0.18		1.1 Fe, 1.0 Ni 0.07 Ti
Aluminium–magnesium–silicon alloys							
6061	97.9		1.0	0.28	0.6	0.2	
6063	98.9		0.7		0.4		
6151	98.2		0.6		0.9	0.25	
6262	96.8		1.0	0.28	0.6	0.09	0.6 Bi, 0.6 Pb
Aluminium–zinc–magnesium–copper alloys							
7075	90.0		2.5	1.6		0.23	5.6 Zn
7178	88.1		2.7	2.0		0.26	6.8 Zn

4.3 Heat treatment

Annealing

Typical annealing temperatures and procedures for Aluminium Association specification wrought alloys to bring them to the O temper are as follows. For the 1XXX, 3XXX, 5XXX series a temperature of 345°C with both the soaking time and cooling rate not being important. 3003 is an exception to this, requiring a temperature of 415°C. The 2XXX and 6XXX series require temperatures of 415°C with a soaking time of 2 to 3 hours and a cooling rate of about 30°C/h to 260°C. An exception to this is 2036 with a temperature of 385°C. The 7XXX series requires a temperature of 415°C and a soaking time of 2 to 3 hours. There is then uncontrolled cooling to about 200°C followed by reheating to

230°C for 4 hours. An exception to this is 7005 which requires a temperature of 345°C, a soaking time of 2 to 3 hours followed by a cooling rate of 30°C/h or less to 200°C.

Heat treatment for cast alloys

Table 4.11 shows typical heat treatments, solution heat treatment and aging, for commonly used Aluminium Association cast alloys. The solution heat treatment is generally followed by quenching in water at 65 to 100°C. Table 4.12 shows similar data for British Standard cast alloys.

Table 4.11 Heat treatment for AA cast alloys

AA no.	Cast. type	Solution h.t.		Aging h.t.		Temper
		temp. °C	time h	temp. °C	time h	
222.0	S			315	3	O
	S	510	12	155	11	T61
	P			170	16–22	T551
242.0	S			345	3	O
	S			205	8	T571
	S/P	515	4–12	205–230	3–5	T61
295.0	S	515	12			T4
	S	515	12	155	3–6	T6
	S	515	12	260	4–6	T7
355.0	S/P			225	7–9	T51
	S	525	12	155	3–5	T6
	P	525	4–12	155	2–5	T6
	S	525	12	225	3–5	T7
	P	525	4–12	225	3–9	T7
356.0	S/P			225	7–9	T51
	S	540	12	155	3–5	T6
	P	540	4–12	155	2–5	T6
	S	540	12	205	3–5	T7
	P	540	4–12	225	7–9	T7
357.0	S	540	10–12	155	10–12	T61
	P	540	8	175	6	T6

Note: S = sand casting, P = permanent mould casting.

Table 4.12 Heat treatment for BS cast alloys

BS	Solution h.t.		Quench medium	Precipitation h.t.		Temper
	temp. °C	time h		temp. °C	time h	
LM4	505–520	6–16	W 70–80	150–170	6–18	TF(T10)
LM9				150–170	16	TE(T6)
	520–535	2–8	W	150–170	16	TF(T6)
LM10	425–435	8	O 160m			TB(T4)
LM13				160–180	4–16	TE(T10)
	515–525	8	W 70–80	160–180	4–16	TF(T6)
	515–525	8	W 70–80	200–250	4–16	TF7
LM16	520–530	12	W 70–80			TB(T4)
	520–530	12	W 70–80	160–170	8–10	TF(T6)
LM25	525–545	4–12	W 70–80	250	2–4	TB7(T7)
				155–175	8–12	TE(T10)
	525–545	4–12	W 70–80	155–175	8–12	TF(T6)
LM30				175–225	8	TS(T)

Note: Quench medium; W = water, W 70–80 = water at 70 to 80°C, O 160m = oil at a maximum of 160°C.

Heat treatment for wrought alloys

Table 4.13 gives the solution heat treatment and precipitation heat treatment temperatures, with the resulting tempers for commonly used wrought alloys to the Aluminium Association specification. The tempers produced depend on the product form, e.g. whether it is flat sheet or extruded bar, and any other treatments carried out between solution and precipitation treatments. The table can thus only give an indication of possible outcomes. After solution treatment the alloys are quenched in water at room temperature.

Table 4.13 Solution and precipitation heat treatments

AA	Solution h.t.		Precipitation h.t.		
	temp. °C	temper	temp.° C	time h	temper
Aluminium–copper alloys					
2011	525	T4(TB)			
		+ CW			
		T3(TD)	160	14	T8(TH)
2014	500	T4(TB)	160	10/18	T6(TF)
		+ CW			
		T3(TD)	160	18	T6(TF)
2024	495	T4(TB)			
		+ CW			
		T3(TD)	190	12	+ CW T81(TH)
2219	535	+ CW			
		T31(TD)	190	18	+ CW T81(TH)
		T42(TB)	190	36	T62(TF)
Aluminium–magnesium–silicon alloy					
6061	530	T4(TB)	160	18	T6(TF)
		T42(TB)	160	18	T62(TF)
6063	520	T4(TB)	175	8	T6(TF)
		T42(TB)	175	8	T62(TF)
6151	515	T4(TB)	170	10	T6(TF)
6262	540	T4(TB)	170	8	T6(TF)
		T42(TB)	170	8	T62(TF)
Aluminium–zinc–magnesium–copper alloys					
7075	480	W	120	24	T6(TF)
		+ SR W51	120	24	T651

Note: CW = cold worked, SR = stress relieved. Where possible the nearest equivalent BS temper has been given in parentheses.

4.4 Properties

Density

The density of aluminium at 20°C is 2.69 g cm^{-3} and the alloys have values between about 2.6 and 2.8 g cm^{-3}.

Electrical properties

Tables 4.14 and 4.15 show, for a temperature of 20°C, the electrical resistivities in units of Ω m and electrical conductivities on the IACS scale. This scale expresses conductivities as a percentage of the conductivity of an annealed copper standard at 20°C, this having a resistivity of 1.7241×10^{-8} Ω m.

Table 4.14 Electrical properties of cast aluminium alloys

BS/AA	Temper	Conductivity (% IACS)	Resistivity ($\mu\Omega$ m)
LM4/319.0	M, TF (F, T6)	32	0.054
LM5/514.0	M (F)	31	0.056
LM6	M	37	0.047
LM9/A360.0	M, TE (F, T10)	38	0.045
LM10/520.0	TB (T4)	20	0.086
LM12/222.0	M (F)	33	0.052
LM13/A332.0	TE, TF (T10, T6)	29	0.059
LM16/355.0	TB, TF (T4, T6)	36	0.048
LM18/443.0	M (F)	37	0.047
LM20/413.0	M (F)	37	0.047
LM21/319.0	M (F)	32	0.054
LM24/380.0	M (F)	24	0.072
LM25/356.0	M, TE (F, T10)	39	0.044
LM30/390.0	M, TS (F, T)	20	0.086

Note: The tempers are given in British Standard designations with those in parentheses being American equivalents.

Table 4.15 Electrical properties of wrought aluminium alloys

AA no.	Temper	Conductivity (% IACS)	Resistivity ($\mu\Omega$ m)
Unalloyed aluminium			
1050	O	61	0.028
1060	O	62	0.028
	H18	61	0.028
1100	O	59	0.029
	H18	57	0.030
1200	O, H4, H8	60	0.029
Aluminium–copper alloys			
2011	T3, T4 (TD, TB)	39	0.044
	T8 (TH)	45	0.038
2014	O	50	0.034
	T3, T4 (TD, TB)	34	0.051
	T6 (TF)	40	0.043
2024	O	50	0.034
	T3, T4 (TD, TB)	30	0.057
	T6 (TF)	37	0.046
	T8 (TH)	39	0.044
2219	O	44	0.039
	T31, T37 (TD) T351	28	0.062
	T62, T81 (TF, TH) T851	30	0.057
2618	T61 (TF)	37	0.047
Aluminium–manganese alloys			
3003	O	47	0.037
	H12	42	0.041
	H14	41	0.042
	H18	40	0.043
3004	All	42	0.041
3105	All	45	0.038
Aluminium–magnesium alloys			
5005	O, H38	52	0.033

5050	O, H38	50	0.034
5052	O, H38	35	0.049
5083	All	29	0.060
5086	All	31	0.056
5154	All	32	0.054
5252	All	35	0.049
5454	All	34	0.051
5456	All	29	0.060

Aluminium-magnesium-silicon alloys

6061	O	47	0.037
	T4 (TB)	40	0.043
	T6 (TF)	43	0.040
6063	O	58	0.030
	T6 (TF)	53	0.033
6151	O	54	0.032
	T4 (TB)	42	0.041
	T6 (TF)	45	0.038
6262	T9	44	0.039

Aluminium-zinc-magnesium-copper alloys

7075	O	43	0.040
	T6 (TF)	35	0.049

Note: The tempers are given in American designations with, where there are differences, the British Standard designations in parentheses.

Fabrication properties

Table 4.16 gives a general indication of the machinability and weldability of aluminium alloys. All the ratings are comparative.

Table 4.16 Fabrication properties of aluminium alloys

Alloy AA	Temper AA (BS)	Machinability	Weldability
Unalloyed aluminium			
1050	O, H12	P	G
	H14–16	F	G
1060	O, H12	P	G
	H14–18	F	G
1100	O, H12	P	G
	H14–18	F	G
1200	O, H12	P	G
	H14–18	F	G
Aluminium-copper alloys			
2011	T3,4,8 (TD,TE,TH)	VG	No
2014	O	G	F
	T3,4,6 (TD,TE,TH)	G/VG	F
2024	T3,4,8 (TD,TE,TH)	G/VG	L
2219	T3,8 (TD,TH)	G/VG	G
2618	T61 (TH)	G/VG	L
Aluminium-manganese alloys			
3003	O,H12	P	G
	H14–18	F	G
3004	O,H32	F	G
	H34–38	G	G
3105	O,H12	P	G
	H14–18	F	G
Aluminium-magnesium alloys			
5005	O,H12	P	G

	H14–18	F	G
5050	O	P	G
	H32–38	F/G	G
5052	O,H32	F	G
	H34–38	G	G
5083	O,H111	F	F
	H323	F	F
5086	O,H32	F	F
	H34–38	G	F
5154	O,H32	F	F
	H34–38	G	F
5252	H24	F	G
	H25–28	G	G
5454	O,H32	F	F
	H34	G	F
5456	O	F	F
	H111,321	F	G
Aluminium–magnesium–silicon alloys			
6061	O	F	G
	T4, T6 (TB,TF)	G	G
6063	T1, T4 (TB)	F	G
	T5, T6 (TF)	G	G
6262	T6, T9 (TF)	G/VG	G
Aluminium–zinc–magnesium–copper alloys			
7075	O	F	No
	T6 (TF)	G/VG	No

Note: Machinability; VG = very good, giving an excellent finish,
G = good, giving a good finish, F = fair, giving a satisfactory finish,
P = poor, care is needed to obtain a satisfactory finish. Weldability;
G = good, readily weldable by all techniques, F = fair, weldable well
by most techniques, L = limited weldability, not all techniques
possible, No = not recommended for welding.

Tempers are given on the American system and where the British
system differs the equivalent tempers are given in parentheses.

Fatigue properties

The fatigue strength, at 50×10^6 cycles, of aluminium alloys tends
to be about 0.3 to 0.5 times the tensile strength. The value for a
particular alloy depends on the temper.

Machinability

See Fabrication properties.

Mechanical properties of cast alloys

Tables 4.17 and 4.18 give the mechanical properties of cast
aluminium alloys. The modulus of elasticity of all the alloys is about
71 GPa or GN m^{-2}, with the exception of those alloys with high
percentages of silicon (greater than about 15%) when the modulus
is about 80 to 88 GPa or GN m^{-2}, the higher the percentage of
silicon the higher the modulus.

Table 4.17 Mechanical properties of AA cast aluminium alloys

AA	Process	Temper	Tensile strength (MPa)	Yield stress (MPa)	Elong- ation (%)
208.0	S	F(M)	145	97	2.5

242.0	S	T21	185	125	1
	S	T571	220	205	0.5
	S	T77	205	160	2
	P	T571	275	235	1
	P	T61	325	290	0.5
295.0	S	T4(TB)	220	110	8.5
	S	T6(TF)	250	165	5
308.0	P	F(M)	195	110	2
319.0	S	F(M)	185	125	2
	S	T6(TF)	250	165	2
	P	F(M)	235	130	2.5
	P	T6(TF)	280	185	3
355.0	S	T6(TF)	240	175	3
	S	T7	265	250	0.5
	P	T6(TF)	290	190	4
	P	T7	280	210	2
356.0	S	T6(TF)	230	165	3.5
	S	T7	235	210	2
	P	T6(TF)	265	185	5
	P	T7	220	165	6
360.0	D	F(M)	325	170	3
A360.0	D	F(M)	320	165	5
380.0	D	F(M)	330	165	3
413.0	D	F(M)	300	140	2.5
C443.0	S	F(M)	130	55	8

Note: S = sand casting, P = permanent mould casting, D = die casting. The values quoted for the yield stress are the 0.2% proof stress. The temper values quoted are American designations with those in parentheses British Standard designations.

Table 4.18 Mechanical properties of BS cast aluminium alloys

BS	Process	Temper	Tensile strength (MPa)	Elong-ation (%)
LM4	S	M(F)	140	2
	S	TF(T6)	230	
	C	M(F)	160	2
	C	TF(T6)	280	
LM5	S	M(F)	140	3
	C	M(F)	170	5
LM6	S	M(F)	160	5
	C	M(F)	190	7
LM9	S	TE(T10)	170	1.5
	S	TF(T6)	240	
	C	M(F)	190	3
	C	TE(T10)	230	2
	C	TF(T6)	295	
LM10	S	TB(T4)	280	8
	C	TB(T4)	310	12
LM12	C	M(F)	170	
LM13	S	TF(T6)	170	
	S	TF7(T7)	140	
	C	TE(T10)	210	
	C	TF(T6)	280	
	C	TF7	200	
LM16	S	TB(T4)	170	2
	S	TF(T6)	230	

	C	TB(T4)	230	3
	C	TF(T6)	280	
LM18	S	M(F)	120	3
	C	M(F)	140	4
LM20	C	M(F)	190	5
LM21	S	M(F)	150	1
	C	M(F)	170	1
LM24	C	M(F)	180	1.5
LM25	S	M(F)	130	2
	S	TE(T10)	150	1
	S	TB7(T7)	160	2.5
	S	TF(T6)	230	
	C	M(F)	160	3
	C	TE(T10)	190	2
	C	TB7(T7)	230	5
	C	TF(T6)	280	2
LM30	C	M(F)	150	

Note: S = sand casting, C = chill casting where a higher rate of cooling occurs than with sand casting, e.g. as with die casting. The tempers are given to British Standard designations with those in parentheses being American designations.

Mechanical properties of wrought alloys

Table 4.19 shows typical mechanical properties of wrought Aluminium Association alloys. The properties will vary to some extent according to the form of the material, e.g. sheet or forging, so the properties must only be regarded as indicative of what might obtain. The modulus of elasticity for all the wrought alloys is about 69 to 74 GPa or GN m^{-2}.

Table 4.19 Mechanical properties of wrought alloys

AA no.	Temper	Tensile strength (MPa)	Yield stress (MPa)	Elong- ation (%)	Hard- ness (HB)
Unalloyed aluminium					
1050	O	76	28		
	H14	110	105		
	H18	160	145		
1060	O	69	28	43	19
	H12	83	76	16	23
	H18	130	125	6	35
1100	O	90	34	35	23
	H12	110	105	12	28
	H18	165	150	5	44
1200	O	87		30	
	H12	108		8	
	H18	150		4	
Aluminium–copper alloys					
2011	T3 (TD)	380	295	15*	95
	T8 (TH)	405	310	15*	100
2014	O	185	97	18*	45
	T4 (TB)	425	290	20*	105
	T6 (TF)	485	415	13*	135
2014Clad	O	170	69	21	
	T4 (TB)	420	255	22	
	T6 (TF)	470	415	10	
2024	O	185	76	20	47

	T3 (TD)	485	345	18	120
	T4 (TB)	470	325	20	120
2024Clad	O	180	76	20	
	T4 (TB)	440	290	19	
	T81(TH)	450	415	6	
2219	O	170	76	18	
	T42(TB)	360	185	20	
	T62(TF)	415	290	10	
	T81(TH)	455	350	10	
2618	All	440	370	10	

Aluminium-manganese alloys

3003	O	110	42	30	28
	H12	130	125	10	35
	H18	200	185	4	55
3004	O	180	69	20	45
	H32	215	170	10	52
	H38	285	250	5	77
3105	O	115	55	24	
	H12	150	130	7	
	H18	215	195	3	

Aluminium-magnesium alloys

5005	O	125	41	25	28
	H12	140	130	10	
	H18	200	195	4	
	H34	160	140	8	41
5050	O	145	55	24	36
	H32	170	145	9	46
	H38	220	200	5	63
5083	O	290	145	22*	
	H112	305	195	16*	
	H34	345	285	9*	
5086	O	260	115	22	
	H32	290	205	12	
	H34	325	244	10	
5154	O	240	115	27	
	H32	270	205	15	67
	H38	330	270	10	80
5252	H25	235	170	11	68
	H28	285	240	5	75
5454	O	250	115	22	62
	H32	275	205	10	73
	H38	370	310	8	
5456	O	310	160	24*	
	H111	325	230	18*	
	H321	350	285	16*	90

Aluminium-magnesium-silicon alloys

6061	O	125	55	25	30
	T4 (TB)	240	145	22	65
	T6 (TF)	310	275	12	95
6061Clad	O	115	48	25	
	T4 (TB)	230	130	22	
	T6 (TF)	290	255	12	
6063	O	90	48	25	
	T4 (TB)	170	90	22	
	T6 (TF)	240	215	12	73
6151	T6 (TF)	220	195	15	71
6262	T9	400	380	10*	120

Aluminium-zinc-magnesium-copper alloys

7075	O	230	105	17	60
	T6 (TF)	570	505	11	150

7075Clad	O	220	95	17	
	T6 (TF)	525	460	11	

Note: * Indicates that elongation is measured for a test piece of 12.5 mm thickness. For all others the test piece thickness is 1.6 mm. The hardness is measured using a 500 kg load with a 10 mm ball. Tempers are designated according to the American system, though where the British Standard system differs the British designation is given in parentheses.

Specific gravity

The specific gravity of aluminium at 20°C is 2.69 and the alloys have values between about 2.6 and 2.8.

Thermal properties

The linear thermal expansivity, i.e. the coefficient of linear expansion, of wrought aluminium alloys tends to be about 22 to 24×10^{-6} °C^{-1}, for cast aluminium alloys it is generally lower at about 18 to 22×10^{-6} °C^{-1}. Both values refer to temperatures in the region 20 to 100°C. The thermal conductivities of wrought alloys at 20°C tends to be between about 100 and 200 W m^{-1} °C^{-1}, for cast alloys between about 90 and 150 W m^{-1} °C^{-1}. Table 4.20 shows values of these properties for some wrought alloys.

Table 4.20 Thermal properties of wrought aluminium alloys

Alloy	Temper	Coeff. of exp. $(10^{-6}\,°C^{-1})$	Thermal conduct. (W m^{-1} °C^{-1})
Unalloyed aluminium			
1050	O, H8	24	230
1060	O, H18	24	234
1100	O	24	222
	H18	24	218
1200	O, H4, H8	24	226
Aluminium–copper alloys			
2011	T3, T6 (TD, TF)	23	163
2014	T4 (TF)	22	142
	T6 (TF)	22	159
2024	T3, T4 (TD, TB)	23	121
2618	T6 (TF)	22	151
Aluminium–manganese alloys			
3003	O	23	180
	H18	23	155
3105	all	24	172
Aluminium–magnesium alloys			
5005	all	24	201
5050	all	24	191
5052	all	24	137
5083	all	25	109
5086	all	24	127
5154	all	25	138
5454	all	24	147
5456	all	24	116
Aluminium–magnesium–silicon alloys			
6061	all	24	156
6063	T4 (TB)	24	197
	T6 (TF)	24	201
6151	T4 (TB)	23	163

T6 (TF)	23		175
Aluminium–zinc–magnesium–copper alloys			
7075 T6 (TF)	24		130

Note: Tempers are given in the American system and where the British system differs the British values are given in parentheses.

Weldability

See Fabrication properties.

4.5 Uses

Forms of material

Table 4.21 shows the normal product forms of wrought material to the Aluminium Association specification.

Table 4.21 Forms of wrought alloy

AA no.	Sheet	Plate	Extruded r.b.w.	shape	tube	Cold finished r.b.w.	Drawn tube	Forgings
Unalloyed aluminium								
1050				*	*	*	*	*
1060				*	*	*	*	
1100	*		*	*	*	* *	*	*
1200	*		*	*	*	* *	*	*
Aluminium–copper alloys								
2011	*		*		*	*	*	
2014	*		*	*	*	* *	*	*
2024	*		*	*	*	* *	*	
2219	*		*	*	*	* *		*
2618								*
Aluminium–manganese alloys								
3003	*		*	*	*	* *	*	*
3004				*	*	*	*	
3105				*				
Aluminium–magnesium alloys								
5005				*		* *		
5050				*		* *	*	
5052				*		* *	*	
5083	*		*	*	*	*		*
5086	*		*	*	*	*	*	
5154	*		*	*	*	* *	*	
5252				*				
5454	*		*	*	*	*		*
5456	*		*	*	*	*	*	*
Aluminium–magnesium–silicon alloys								
6061	*		*	*	*	* *	*	*
6063	*		*	*	*		*	
6151								*
6262	*		*		*	*	*	
Aluminium–zinc–magnesium–copper alloys								
7075	*		*	*	*	* *	*	*
7178				*		*		*

Uses of cast alloys

Table 4.22 shows typical uses of cast aluminium alloys. The aluminium–silicon alloys with some copper and/or magnesium are very widely used. They have the advantages over the aluminium–

copper alloys of better fluidity and resistance to corrosion.

Table 4.22 Typical uses of cast alloys

Alloy		Uses
Aluminium–copper alloys		
208.0		A general purpose sand casting alloy, used for manifold and valve bodies.
213.0		Used for car cylinder heads, washing machine agitators.
222.0	LM12	Used for pistons.
242.0		Used for pistons in high-performance engines.
295.0		Used for castings requiring high strength and shock resistance.
Aluminium–silicon–copper/magnesium alloys		
308.0		A general purpose permanent mould casting alloy.
319.0	LM4/21	A general purpose alloy, used for engine parts.
336.0	LM13	A permanent mould casting alloy.
355.0	LM16	Widely used where high strength and pressure tightness are required. Uses include pump bodies, crankcases, blower housings.
356.0	LM25	Used for intricate castings requiring strength and ductility. Uses include transmission cases, truck wheels, cylinder blocks, outboard motor parts, fan blades, pneumatic tools.
360.0	LM9	A general purpose die casting alloy. Used for instrument cases.
380.0	LM24	A die casting alloy.
390.0	LM30	A die casting alloy
Aluminium–silicon alloys		
413.0	LM20	A die casting alloy used for large intricate castings with thin sections, e.g. typewriter frames.
C443.0	LM18	A die casting alloy used for castings requiring high resistance to corrosion and shock.
Aluminium–magnesium alloys		
514.0	LM5	A sand casting alloy
520.0	LM10	A sand casting alloy

Uses of wrought alloys

Table 4.23 shows typical uses of wrought aluminium alloys.

Table 4.23 Typical uses of wrought aluminium alloys

Alloy	Uses
Unalloyed aluminium	
1050	Extruded coiled tubing, chemical equipment.
1060	Chemical equipment.
1100	Sheet metal work, spun hollow-ware.
1200	Extruded coiled tubing, sheet metal work.
Aluminium–copper alloys	
2011	Screw machine products.
2014	Aircraft structures, frames for trucks.

2024	Aircraft structures, truck wheels.
2219	Used for high strength welds in structures at temperatures as high as 315°C, e.g. aircraft parts.
2618	Aircraft engine parts.

Aluminium–manganese alloys

3003	Has a wide general use, e.g. cooking utensils, sheet metal work, builders' hardware, storage tanks, pressure vessels, chemical equipment.
3004	Sheet metal work, storage tanks.
3105	Builders' hardware, sheet metal work.

Aluminium–magnesium alloys

5005	Electrical conductors, architectural trims, general utensils.
5050	Builders' hardware, coiled tubes.
5052	Sheet metal work, hydraulic tubes.
5083	Welded pressure vessels, marine, car and aircraft parts.
5086	As 5083.
5154	Welded structures, storage tanks, pressure vessels.
5252	Car and appliance trims.
5454	Welded structures, pressure vessels, marine applications.
5456	Where high strength, welded structures are required. Storage vessels, pressure vessels, marine applications.

Aluminium–magnesium–silicon alloys

6061	Heavy duty structures where good corrosion resistance is required, e.g. truck and marine applications, pipelines, furniture.
6063	Architectural extrusions, pipes, furniture.
6151	Where moderate strength, intricate forgings are required, e.g. car and machine parts.
6262	Screw machine products.

Aluminium–zinc–magnesium–copper alloys

| 7075 | Hydraulic fittings, aircraft structures. |
| 7178 | As 7075. |

5 Copper

5.1 Materials

Copper

Copper has very high electrical and thermal conductivity and can be manipulated readily by either hot or cold working. Pure copper is very ductile and relatively weak. Working increases the tensile strength and hardness but decreases ductility. It has a good corrosion resistance, due to reactions at the surface between the copper and the oxygen in air to give a thin protective oxide layer.

Copper is widely used for electrical conductors in a number of high purity grades and with the addition of very small amounts of arsenic, phosphorus, silver, sulphur or tellurium. Very pure copper can be produced by an electrolytic refining process, the pure copper forming at the cathode of the electrolytic cell and so being referred to as cathode copper. It has a purity greater than 99.99%, being mainly used as the raw material for the production of alloys, with some use also being made of it as a casting material. Electrolytic tough pitch high-conductivity copper is produced from cathode copper which has been melted and cast into billets, and other suitable shapes, for working. It contains a small amount of oxygen, present in the form of cuprous oxide, which has little effect on the electrical conductivity. This type of copper should not be heated in an atmosphere where it can combine with hydrogen, because the hydrogen can diffuse into the metal and combine with the cuprous oxide to generate steam and hence crack the copper. Fire refined tough pitch high-conductivity copper is produced from impure copper by a fire refining process in which the impure copper is melted in an oxidizing atmosphere. The impurities react with the oxygen to give a slag which is removed. The resulting copper has an electrical conductivity almost as good as the electrolytic tough high-pitch high-conductivity copper. Oxygen-free high-conductivity copper can be produced if, when cathode copper is melted and cast into billets, there is no oxygen present in the atmosphere. Such copper can be used in atmospheres containing hydrogen. Another method of producing oxygen-free copper is to add phosphorus during the refining. The effect of small amounts of phosphorus is a very marked decrease in electrical conductivity. Such copper is known as phosphorus deoxidized copper and it can give good welds, unlike other forms of copper. The addition of small amounts of arsenic to copper increases its tensile strength, but greatly reduces its electrical conductivity. Such copper is called arsenical copper.

See Codes for composition, Codes for temper, Composition of casting alloys, Composition of wrought alloys, Annealing, Creep properties, Hardness, Mechanical properties of cast alloys, Mechanical properties of wrought alloys, Solderability, Thermal properties, Weldability, Forms, Uses of cast alloys, Uses of wrought alloys.

Copper alloys

Table 5.1 shows the main groups of copper alloys and their main alloying elements. The relevant portions of the binary alloy equilibrium diagrams are given in Figures 5.1, 5.2, 5.3, 5.4, 5.5 and 5.6. Many of the alloys within alloy groups are given specific names.

Table 5.1 Main groups of copper alloys

Group name	Main alloying elements	Equilibrium diagram
Brasses	copper, zinc	Figure 5.1
Bronzes	copper, tin	Figure 5.2
Phosphor bronzes	copper, tin, phosphorus	
Gun metals*	copper, tin, zinc	
Aluminium bronzes	copper, aluminium	Figure 5.3
Silicon bronzes	copper, silicon	Figure 5.4
Beryllium bronzes	copper, beryllium	Figure 5.5
Cupro-nickels	copper, nickel	Figure 5.6
Nickel-silvers	copper, nickel, zinc	

* In America these are generally referred to as tin brasses.

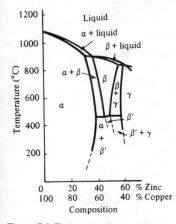

Figure 5.1 The copper-zinc equilibrium diagram

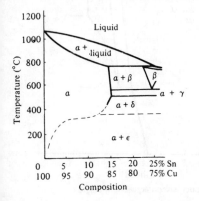

Figure 5.2 The copper-tin equilibrium diagram

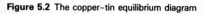

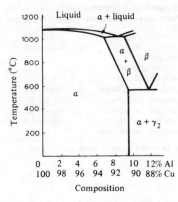

Figure 5.3 The copper–aluminium equilibrium diagram

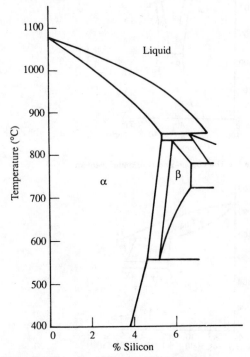

Figure 5.4 The copper–silicon equilibrium diagram

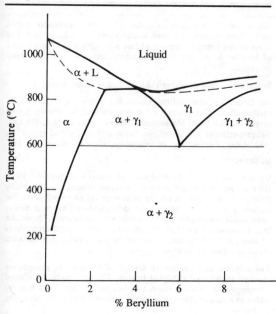

Figure 5.5 The copper–beryllium equilibrium diagram

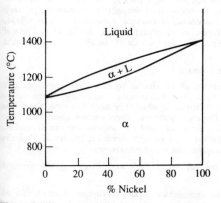

Figure 5.6 The copper–nickel equilibrium diagram.

Brasses

The brasses are copper–zinc alloys containing up to about 43% zinc. Brasses with less than 35% zinc solidify as single phase alpha brasses. These brasses have high ductility and can readily be cold worked. A brass with 15% zinc is known as gilding brass, one with 30% zinc as cartridge brass or 70/30 brass.

At between 35 and 45% zinc, the brasses solidify as a mixture of two phases. These brasses are easier to hot work but harder to cold work. These brasses are known as alpha-beta or duplex brasses. The name Muntz metal is given to a brass with 40% zinc. The addition of lead to Muntz metal considerably improves the machining properties.

Brasses are available in both casting and wrought forms.

See Codes for composition, Codes for temper, Composition of casting alloys, Composition of wrought alloys, Annealing, Creep properties, Hardness, Mechanical properties of cast alloys, Mechanical properties of wrought alloys, Solderability, Thermal properties, Weldability, Forms, Uses of cast alloys, Uses of wrought alloys.

Bronzes

Up to about 10% tin the structure that occurs with normal cooling from the liquid state is a single phase alpha structure, such a phase giving a ductile material. Higher percentage tin alloys will invariable include a significant amount of delta phase. This is a brittle intermetallic compound. Alpha bronzes with up to about 8% tin can be cold worked to give high mechanical properties. High tin contents of about 10%, or higher, make the bronze unworkable but such alloys are used for casting.

The addition to a copper–tin bronze of up to 0.4% phosphorus gives an alloy called phosphor bronze. The term gunmetal is used for copper–tin bronzes when zinc is also present. Lead may be added to a gunmetal to give a leaded gunmetal with good machining properties.

Bronzes are available in both casting and wrought forms.

See Codes for composition, Codes for temper, Composition of casting alloys, Composition of wrought alloys, Annealing, Creep properties, Hardness, Mechanical properties of cast alloys, Mechanical properties of wrought alloys, Solderability, Thermal properties, Weldability, Forms, Uses of cast alloys, Uses of wrought alloys.

Aluminium bronzes

Copper-aluminium alloys with up to about 9% aluminium give single phase alpha aluminium bronzes. These alloys can be readily cold worked, particularly when they have less than about 7% aluminium. Duplex, alpha-beta, alloys contain about 9 to 10% aluminium and are mainly used for castings. The cast aluminium bronzes, and many of the wrought aluminium bronzes, also contain iron and sometimes nickel and manganese.

Aluminium bronzes are available in both casting and wrought forms.

See Codes for composition, Codes for temper, Composition of casting alloys, Composition of wrought alloys, Annealing, Quenching and tempering, Creep properties, Hardness, Mechanical properties of cast alloys, Mechanical properties of wrought alloys, Solderability, Thermal properties, Weldability, Forms, Uses of cast alloys, Uses of wrought alloys.

Silicon bronzes

Silicon bronzes are copper–silicon alloys, generally containing between about 1 and 4% silicon and giving a single phase alpha structure. The alloys have a high corrosion resistance, very good weldability and can be cast or hot or cold worked.

See Codes for composition, Codes for temper, Composition of wrought alloys, Annealing, Creep properties, Mechanical properties of wrought alloys, Solderability, Thermal properties, Weldability, Forms, Uses of wrought alloys.

Beryllium bronzes

There are two groups of beryllium bronzes, one with about 0.4% beryllium and the other with about 1.7 to 2.0% beryllium. Cobalt is also usually present. The alloys have high strength and hardness, but are expensive.

See Codes for composition, Codes for temper, Composition of wrought alloys, Annealing, Precipitation hardening, Creep properties, Hardness, Mechanical properties of wrought alloys, Solderability, Thermal properties, Weldability, Forms, Uses of wrought alloys.

Cupro-nickels

Copper and nickel are completely soluble in each other in both the liquid and solid states. They thus form a single phase alpha structure over the entire range of compositions from 0 to 100%. They may be cold or hot worked over the entire range. The alloys have high strength and ductility, and good corrosion resistance.

If zinc is added, the resulting alloy is called a nickel silver, the alloy having a silver colour. These alloys tend to have about 8 to 18% nickel and 17 to 27% zinc. They have good cold formability and corrosion resistance. The alloys are generally single phase alpha structures.

Both cupro-nickels and nickel silvers are mainly used in wrought forms, though there is some use made of cast nickel silvers.

See Codes for composition, Codes for temper, Composition of cast alloys, Composition of wrought alloys, Annealing, Creep properties, Hardness, Mechanical properties of cast alloys, Mechanical properties of wrought alloys, Solderability, Thermal properties, Weldability, Forms, Uses of cast alloys, Uses of wrought alloys.

5.2 Codes and compositions

Codes for composition

A commonly used American system is that of the Copper Development Association (CDA) and is shown in Table 5.2. This uses the letter C followed by three digits. The first digit indicates the group of alloys concerned and the remaining two digits alloys within the group. The groups C1XX to C7XX are used for wrought alloys and C8XX with C9XX for cast alloys. Frequently in references the letter C is not included, just the three digits being used.

In the British Standards (BS) system, wrought copper and copper alloys are specified by two letters followed by three digits. The two letters indicate the alloy group and the three digits alloys within that group. Table 5.3 describes the system. Letters followed by a digit are used to describe casting alloys, Table 5.4 describing the system used.

Table 5.2 CDA codes for copper and its alloys

CDA code	Alloy group
Wrought alloys	
C1XX	Coppers with a minimum of 99.3% Cu, high-copper alloys with between 96 and 99.3% Cu.
C2XX	Copper-zinc alloys, i.e. brasses.
C3XX	Copper-zinc-lead alloys, i.e. leaded brasses.
C4XX	Copper-zinc-tin alloys, i.e. tin brasses.
C5XX	Copper-tin alloys, i.e. bronzes/phosphor bronzes.
C6XX	Copper-aluminium alloys. i.e. aluminium bronzes, Copper-silicon alloys, i.e. silicon bronzes, and miscellaneous copper-zinc alloys.
C7XX	Copper-nickel alloys, i.e. cupro-nickels, Copper-nickel-zinc alloys, i.e. nickel silvers.
Cast alloys	
C8XX	Cast coppers, high-copper alloys, brasses of various types, manganese-bronze alloys, copper-zinc-silicon alloys.
C9XX	Cast copper-tin alloys, copper-tin-lead alloys, copper-tin-nickel alloys, copper-aluminium-iron alloys, copper-nickel-iron alloys, copper-nickel-zinc alloys.

Table 5.3 BS Codes for wrought copper and its alloys

BS code	Alloy group
C	Copper and low-copper alloys.
CA	Copper-aluminium alloys, i.e. aluminium bronzes.
CB	Copper-beryllium alloys, i.e. beryllium bronzes.
CN	Copper-nickel alloys, i.e. cupro-nickels.
CS	Copper-silicon alloys, i.e. silicon bronzes.
CZ	Copper-zinc alloys, i.e. brasses.
NS	Copper-zinc-nickel alloys, i.e. nickel silvers.
PB	Copper-tin-phosphorus, i.e. phosphor bronzes.

Table 5.4 BS Codes for cast copper and its alloys

BS code	Alloy
AB1, AB2, AB3	Aluminium bronze
CMA1	Copper-manganese-aluminium
CN1	Copper-nickel-chromium
CN2	Copper-nickel-niobium
CT1	Tin bronze for general purposes
CT2	Tin bronze with nickel
DCB1, DCB3	Brass for gravity die casting
G1	80/10/2 Admiralty gunmetal
G3, G3-TF	Nickel gunmetal
HCC1	High conductivity copper
HTB1, HTB3	High tensile brass
LB1	79/9/0/15 Leaded bronze
LB2	80/10/0/10 Leaded bronze
LB4	85/5/0/10 Leaded bronze
LB5	75/5/0/20 Leaded bronze
LG1	83/3/9/5 Leaded gunmetal
LG2	85/5/5/5 Leaded gunmetal

LG4	87/7/3/3 Leaded gunmetal
LPB1	Leaded phosphor bronze
PB1, PB4	Phosphor bronze for bearings
PB2	Phosphor bronze for gears
PCB1	Brass for pressure die casting
SCB1, SCB3	Brass for general purposes
SCB4	Naval brass for sand casting
SCB6	Brazable quality brass

Codes for tempers

The American Society for Metals have over a hundred designations of temper for copper alloys. Considerably fewer designations are used with the British Standards' system. Table 5.5 shows the basis of both systems and their interrelationship.

Table 5.5 ASTM and BS codes for tempers

BS		ASTM	
Code	Temper condition	Code	Temper condition
O	Annealed	O10–O82	Annealed to produce specific properties, e.g.
		O50	Light annealed
		O70	Dead soft annealed
		OS005–OS200	Annealed to produce specific grain size, e.g.
		OS005	Average size 0.005 mm
		OS200	Average size 0.200 mm
H	Cold worked to hardnesses, e.g.	H00–H14	Cold worked to hardnesses, e.g.
		H00	One-eighth hard
¼H	Quarter hard	H01	One-quarter hard
½H	Half hard	H02	Half hard
		H03	Three-quarter hard
H	Hard	H04	Hard
EH	Extra hard	H06	Extra hard
SH	Spring hard	H08	Spring hard
ESH	Extra spring hard	H10	Extra spring
		H12	Special spring
		H13	Ultra spring
		H14	Super spring
		H50–H86	Cold worked via specific processes, e.g.
		H50	Extruded and drawn
		HR01–HR50	Cold worked and stress relieved, e.g.
		HR01	H01 and stress relieved.
		HT04–HT08	Cold worked and order strengthened, e.g.
		HT04	H04 and order heat treated
M	As manufactured.	M01–M45	As manufactured, e.g.
		M01	As sand cast
		M02	As centrifugal cast
		M20	As hot rolled
		M30	As hot extruded
W	Solution treated.	TB00	Solution heat treated

W(xH)	Sol.tr. and cold worked to temper	TD00–TD04	Sol.tr. and cold worked to temper
		TD00	TB00 to 1/8 hard
W(¼H)	To 1/4 hard	TD01	TB00 to 1/4 hard
W(½H)	To 1/2 hard	TD02	TB00 to 1/2 hard
		TD03	TB00 to 3/4 hard
W(H)	To full hard	TD04	TB00 to full hard
—	—	—	—
WP	Sol.tr. and prec. hardened	TF00	TB00 and precipitation hardened
—	—	—	—
W(xH)P	Sol.tr., c.w. and prec. hardened	TL01–TL04	Sol.tr., cold worked and prec. hardened
		TL00	TF00 c.w. to 1/8 hard
W(¼H)P	C.w. to 1/4 hard	TL01	TF00 c.w. to 1/4 hard
W(½H)P	C.w. to 1/2 hard	TL02	TF00 c.w. to 1/2 hard
W(H)P	C.w. to full hard	TL04	TF00 c.w. to full hard
		TR01–TR04	TL and stress relieved
		TR01	TL01 and stress relieved
		TR02	TL02 and stress relieved
		TR04	TL04 and stress relieved
		TH01–TH04	TD and prec. hardened
		TH01	TD01 and prec. hardened
		TH02	TD02 and prec. hardened
		TH03	TD03 and prec. hardened
		TH04	TD04 and prec. hardened
Wm	Mill hardened	TM00–TM08	Mill hardened to various tempers.
		TQ00–TQ75	Quench hardened tempers
		TQ00	Quench hardened
		TQ50	Quench hardened and temper annealed
		TQ75	Interrupted quench hardened
			Tempers of welded tubes
		WH00–WH01	Welded and drawn
		WM00–WM50	As welded from Hxx strip.
		W050	Welded and light anneal
		WR00–WR01	WM00/01 drawn and stress relieved.

Composition of casting alloys

Table 5.6 shows the composition of commonly used British Standards and Table 5.7 the Copper Development Association specified casting alloys. Where there is a common name for an alloy it is given in addition to the BS or CDA code. In British Standards casting alloys are grouped into three categories:

Group A: Alloys in common use and preferred for general purposes.

Group B: Special purpose alloys for which particular properties are required.

Group C: Alloys in limited production.

These categories are indicated in the table for those alloys to British Standards.

Table 5.6 Composition of BS casting alloys

Name	BS code	Group	Nominal Composition %

Copper			Cu			
High conductivity	HCC1	B	>99.9			
Brasses			Cu	Sn	Zn	Other
Sand cast brass	SCB1	A	70–77	1–3	rem	2–5 Pb
Sand cast brass	SCB3	A	63–66		rem	1–3 Pb
Sand cast naval brass	SCB4	C	60–63	1–1.5	rem	
Brazable sand cast brass	SCB6	A	83–88		rem	0.05–0.2 As
Gravity die cast brass	DCB1	A	59–62		rem	0.25–0.5 Al
Gravity die cast naval	DCB3	A	58–62		rem	0.5–2.5 Pb, 0.2–0.8 Al
Pressure die cast brass	PCB1	A	57–60		rem	0.5–2.5 Pb
High tensile brass	HTB1	B	>57.0		rem	0.7–2.0 Fe, 0.1–3.0 Mn, 0.5–2.5 Al
High tensile brass	HTB3	B	>55.0		rem	1.0 Ni, 1.5–4.0 Mn, 1.5–3.25 Fe, 3.0–6.0 Al

Bronzes			Cu	Sn	Other	
Tin bronze	CT1	B	90	10	0.05 P	
Phosphor bronze	PB1	B	89	10–11	0.6–1.0 P	
Phosphor bronze	PB2	B	88	11–13	0.25–0.60 P	
Phosphor bronze	PB4	A	89	>9.7	>0.5 P	
Leaded phosphor bronze	LPB1	A	87	7.5	3 Pb, >0.4 P, 1.0 Ni	
Leaded br. 76-9-0-15	LB1	C	76	9	15 Pb, 0.50 Sb	
Leaded br. 80-10-0-10	LB2	A	80	10	10 Pb, 0.50 Sb	
Leaded br. 85-5-0-10	LB4	A	85	5	10 Pb, 0.50 Sb	
Leaded br. 75-5-0-20	LB5	C	75	5	20 Pb, 0.50 Sb	

Gunmetals			Cu	Sn	Zn	Pb	Other
Admiralty gunmetal	G1	C	88	10	2		
Nickel gunmetal	G3	C	86	7	2	0.3	5.5 Ni
Leaded g.m 83-3-9-5	LG1	B	83	3	9	5	
Leaded g.m 85-5-5-5	LG2	A	85	5	5	5	
Leaded g.m 87-7-3-3	LG4	A	87	7	3	3	

Aluminium bronzes			Cu	Al	Fe	Ni	Mn
Aluminium bronze	AB1	B	88	9.5	2.5		
Aluminium bronze	AB2	B	80	9.5	5	5	
Cu-Mn-Al	CMA1	B	73	9	3	3	13

Nickel-silvers			Cu	Ni	Fe	Mn	Nb

Cu–Ni–Cr	CN1	C	67	31	0.7	0.8
Cu–Ni–Nb	CN2	C	66	30	1.2	1.3 1.3

Note: rem = remainder. For clarification of the group symbols see the note preceding this table.

Table 5.7 Composition of CDA casting alloys

Name	CDA	Nominal composition %				
Copper		Cu			Other	
	C801	>99.95			Ag	
	C811	>99.70			Ag	
High copper alloys		Cu	Co	Ni	Other	
	C817	>94.25	0.9	0.9	1 Ag, Be	
	C821	97.7	0.9	0.9	0.5 Be	
Brasses		Cu	Sn	Zn	Pb	Other
Leaded red brass	C836	85	5	5	5	
Leaded red brass	C838	83	4	7	6	
Leaded semi-red brass	C844	81	3	9	7	
Leaded semi-red brass	C848	76	3	15	6	
Leaded yellow brass	C852	72	1	24	3	
Leaded yellow brass	C854	67	1	29	3	
Leaded naval brass	C857	63	1	35	1	0.3 Al
Leaded yellow brass	C858	58	1	40	1	
Manganese bronzes		Cu	Zn	Fe	Al	Mn	Pb
Manganese bronze	C862	64	26	3	4	3
Manganese bronze	C863	63	25	3	6	3
Tin bronzes		Cu	Sn	Zn	Pb	Ni
Tin bronze	C903	88	8	4		
Tin bronze	C905	88	10	2		
Tin bronze	C911	88	14	1		
Steam bronze/valve bronze	C922	88	6	4.5	1.5	
Leaded tin bronze	C923	87	8	4	1	
High-lead tin bronze	C932	83	7	3	7	
High-lead tin bronze	C937	80	10	[ ]	10	
High-lead tin bronze	C938	78	7		15	
High-lead tin bronze	C943	70	5		25	
Nickel-tin bronze	C947	88	5	2		5
Aluminium bronzes		Cu	Al	Fe	Ni	Mn
Aluminium bronze	C952	88	9	3		
Aluminium bronze	C953	89	10	1		
Nickel aluminium bronze	C955	81	11	4	4	
Nickel aluminium bronze	C958	81	9	4	5	1
Silicon bronzes/brasses		Cu	Zn	Si		
Silicon brass	C875	82	14	4		
Nickel-silvers		Cu	Sn	Zn	Pb	Ni
Nickel-silver	C973	56	2	20	10	12
Nickel-silver	C976	64	4	8	4	20
Nickel-silver	C978	66	5	2	2	25

Composition of wrought alloys

Tables 5.8 and 5.9 show the compositions of commonly used British Standard and Copper Development Association wrought alloys.

Table 5.8 Composition of BS wrought alloys

Name	BS	Nominal composition %	
Copper		Cu	Other
Elec. tough pitch h.c. Cu	C101	>99.90	O_2

	Cu	Other
Oxygen free h.c. copper	C103	>99.95
Tough pitch arsenical Cu	C105	>99.20 As
Phosphorus deoxidized Cu	C106	>99.85 P
Phosphorus deox. ars. Cu	C107	>99.20 As, P
Oxygen free h.c. copper	C110	>99.99

High copper alloys

	Cu	Other
Copper–cadmium	C108	99.0	1.0 Cd

Brasses

		Cu	Sn	Zn	Al	Fe	Mn	Oth.
Gilding metal (90/10 br.)	CZ101	90		10				
Red brass (85/15 br.)	CZ102	85		15				
70/30 arsenical brass	CZ105	72		28				As
Deep drawing brass (70/30)	CZ106	70		30				
Basis or common brass	CZ108	63		37				
Muntz metal (60/40 br.)	CZ109	60		40				
Aluminium brass	CZ110	76		22	2			As
Admiralty brass	CZ111	70	1	29				As
Naval brass	CZ112	62	1	37				
High tensile brass	CZ114	58	0.6	37	<1.5	0.85	1.5	Pb
High tensile soldering br.	CZ115	57.5	0.9	38	<0.2	0.85	1.5	Pb
High tensile brass	CZ116	66		27	4.5	0.75	1.5	
Leaded brass (clock br.)	CZ118	64		35				Pb
Leaded Muntz metal	CZ123	60		39				Pb

Phosphor bronzes

		Cu	Sn	P
4% Phosphor bronze	PB101	96	4.0	0.02-0.40
5% Phosphor bronze	PB102	95	5.0	0.02-0.40
7% Phosphor bronze	PB103	93	7.0	0.02-0.40

Aluminium bronzes

		Cu	Al	Fe	Ni	Mn
7% Aluminium bronze	CA102	93	7			
9% Aluminium bronze	CA103	87	9	4.0 + Ni		
10% Aluminium bronze	CA104	80	10	5	5	
Aluminium br.(alloy E)	CA105	81	9	2.0	5.5	1.2
Aluminium br.(alloy D)	CA106	90	7.2	2.7		

Silicon bronze

		Cu	Si	Mn
Copper silicon	CS101	96	2.7-3.5	0.75-1.25

Beryllium bronzes

		Cu	Be	Co + Ni
Copper-beryllium	CB101	98	1.7	0.20-0.60

Cupro-nickels

		Cu	Ni	Fe	Mn
95/5 Cupro-nickel	CN101	93	5.5	1.2	0.5
90/10 Cupro-nickel	CN102	87.5	10.5	1.5	0.75
80/20 Cupro-nickel	CN104	80	20		0.30
75/25 Cupro-nickel	CN105	75	25		0.2
70/30 Cupro-nickel	CN107	68	31		1.0
Special 70/30 Cupro-nick.	CN108	66	30	2.0	2.0

Nickel silvers

		Cu	Ni	Zn	Pb	Mn
Leaded nickel brass	NS101	45	10	43	1-2	0.2-0.5
10% nickel silver	NS103	63	10	27		0.05-0.3
12% nickel silver	NS104	63	12	24		0.05-0.3
15% nickel silver	NS105	63	15	21		0.05-0.5
18% nickel silver	NS106	63	18	19		0.05-0.5
18% nickel silver	NS107	55	18	27		0.05-0.35
20% leaded nickel brass	NS111	60	10	28	1-2	0.1-0.5

Table 5.9 Composition of CDA wrought alloys

Name	CDA	Nominal composition %	

Copper

		Cu	Other
Oxygen free electronic	C101	99.99	
Oxygen free copper	C102	99.95	
Electrolytic tough pitch	C110	99.90	0.04 O_2, 0.01 Cd

		Cu				Other
Phosphorus deoxidized Cu	C122	99.90		0.02 P		
Phosphorus deox. arsenical	C142	99.68		0.3 As, 0.02 P		
Deoxidized cadmium copper	C143	99.90		0.1 Cd		

High copper alloys		Cu		Other		
Cadmium copper	C162	99.0		1.0 Cd		
Chromium copper	C182	99.1		0.9 Cr		

Brasses		Cu	Sn	Zn	Al	Other
Gilding metal	C210	95		5		
Commercial bronze, 90%	C220	90		10		
Red brass	C230	85		15		
Cartridge brass, 70%	C260	70		30		
Yellow brass	C268-0	65		35		
Muntz metal	C280	60		40		
Free-cutting brass	C360	61.5		35.5		3 Pb
Free-cutting Muntz metal	C370	60.0		39		1 Pb
Inhibited admiralty	C443-5	71	1	28		
Naval brass	C464-0	60	0.75	39.25		
Manganese brass	C674	58.5	1	36.5	1.2	2.8 Mn
Aluminium brass, arsenical	C687	77.5		20.5	2.0	0.1 As

Phosphor bronzes		Cu	Sn	P		
Phosphor bronze, 1.25% E	C505	98.75	1.25	trace		
Phosphor bronze, 5% A	C510	95.0	5.0	trace		
Phosphor bronze	C511	95.6	4.2	0.2		
Phosphor bronze, 8%	C521	92.0	8.0	trace		
Phosphor bronze, 10%	C524	90.0	10.0	trace		

Aluminium bronzes		Cu	Al	Fe	Ni	Other
Aluminium bronze, 5%	C608	95	5			
Aluminium bronze	C613	92.65	7.0			0.35 Sn
Aluminium bronze, D	C614	91	7	2		
Aluminium bronze	C623	87	10	3		
Aluminium bronze	C630	82	10	3	5	

Silicon bronze		Cu		Si		
Low silicon bronze B	C651	98.5		1.5		
High silicon bronze, A	C655	97.0		3.0		

Beryllium bronzes		Cu		Be		Co
Beryllium copper	C172	97.9		1.9		0.2

Cupro-nickels		Cu	Ni	Fe		Other
Copper-nickel, 10%	C706	88.7	10	1.3		
Copper-nickel, 20%	C710	79	21			
Copper-nickel, 30%	C715	70	30			
Copper-nickel	C717	67.8	31	0.7		0.5 Be
Copper-nickel	C725	88.2	9.5			2.3 Sn

Nickel silvers		Cu	Ni	Zn		Other
Nickel silver, 65-10	C745	65	10	25		
Nickel silver, 65-18	C752	65	18	17		
Nickel silver, 65-15	C754	65	15	20		
Nickel silver, 65-12	C757	65	12	23		
Nickel silver, 55-18	C770	55	18	27		
Leaded ni. silver, 65-8-2	C782	65	8	25		2 Pb

5.3 Heat treatment

Annealing

Table 5.10 shows typical annealing and stress relieving temperatures for wrought copper and copper alloys. The amount of time at the annealing temperature depends on the amount of prior cold work but is usually 1 to 2 hours. For stress relieving, usually 1 hour at the temperature is required.

Table 5.10 Typical annealing and stress relieving temperatures

Form of alloy	Annealing temp. °C	Stress relieving temp. °C
Copper		
Oxygen free	425–650	
Tough pitch	260–650	
Phosphorus deoxidized	325–650	
Brasses		
Gilding metal	425–800	190
Red brass	425–725	230
Cartridge brass	425–750	260
Yellow or common brass	425–700	260
Muntz metal	425–600	205
Free cutting brass	425–600	245
Phosphor bronzes		
Phosphor bronzes	475–675	205
Aluminium bronzes		
Aluminium bronze, 5%, 7%	550–650	
Aluminium bronze, 10%	600–650	
Al. br. complex alloys	)650	345
Silicon bronzes		
Silicon bronzes	475–700	345
Beryllium bronzes		
Beryllium copper	775–925	
Cupro-nickels		
Cupro-nickels	600–825	260
Nickel silvers		
Nickel silvers	600–825	260

Precipitation hardening

A small number of copper alloys can be hardened and have their strength increased as a result of solution treatment followed by precipitation. These are those alloys that contain small amounts of beryllium, chromium or zirconium, or nickel in combination with silicon or phosphorus. Table 5.11 shows some typical examples.

Table 5.11 Solution treatment and aging treatments

Alloy	Solution tr. temp. °C	Aging tr.	
		temp. °C	time h
Beryllium copper	780–800	300–350	1–3
Chromium copper	980–1000	425–500	2–4
Nickel–tin cast. bronze	775–800	580–620	5

Quenching and tempering

Complex aluminium bronzes, containing more than about 10% aluminium and other elements, can be quenched from about 850°C and then tempered for about 2 hours at 600–650°C . The result is a reduction in the hardness, and increased ductility, from that occurring in the quenched state.

Stress relief

See Annealing.

5.4 Properties

Brazing

See Solderability.

Creep properties

The upper temperature limit of use of copper and its alloys is determined in most instances by creep becoming too pronounced. Table 5.12 shows the upper service temperatures.

Table 5.12 Upper service temperatures for copper alloys

Alloys	Upper service temp. °C
Coppers	120
Brasses, cast	150
Brasses, wrought	180–200
Gunmetals, cast	180
Aluminium bronzes, cast	250–400
Aluminium bronzes, wrought	300–350
Manganese bronzes	200
Phosphor bronzes, wrought	150
Silicon bronze	180
Cupro–nickels, wrought	200
Nickel silvers, wrought	200
Nickel silvers, cast	150

Density

Copper has a density of 8.96×10^3 kg m^{-3} at 20°C. Coppers and high alloy copper have densities of about 8.90×10^3 kg m^{-3}, brasses vary from about 8.40 to 8.90×10^3 kg m^{-3}, e.g. gilding metal 8.86×10^3 kg m^{-3}, yellow brass 8.47×10^3 kg m^{-3}, red brass 8.75×10^3 kg m^{-3}, Muntz metal 8.39×10^3 kg m^{-3}, and Naval brass 8.41×10^3 kg m^{-3}, phosphor bronzes 8.80 to 8.90×10^3 kg m^{-3}, aluminium bronzes about 7.60×10^3 kg m^{-3}, cupro–nickels 8.94×10^3 kg m^{-3}, and nickel silvers about 8.70×10^3 kg m^{-3}.

Electrical conductivity

Copper and high copper alloys have electrical conductivities of the order of 100% IACS, unleaded brasses between about 55 and 25% (the lower the percentage of zinc the higher the conductivity, also the harder the temper the lower the conductivity), leaded brasses about 30 to 25%, phosphor bronzes about 17 to 13%, aluminium bronzes 15 to 7%, silicon bronzes about 7%, cupro–nickels about 12 to 4% and nickel silvers about 10 to 5%.

See the tables with Mechanical properties of cast alloys and Mechanical properties of wrought alloys for specific values of conductivities.

Fatigue properties

The endurance limit of copper and its alloys at about 10^7 cycles is generally between about 0.4 and 0.6 times the tensile strength.

Hardness

The hardness, and other mechanical properties, of a wrought alloy depend on its temper. Table 5.13 shows hardness values, for different tempers, that are typical of the different types of copper alloy. Since the values also depend on the form of the material, the data refer to just sheet or strip.

Table 5.13 Hardness values for different tempers

Alloy	Hardness (HV) values for tempers			
	M	O	½H H02	H H04
Coppers	50	60	80	90
Brasses		80	110	130
Phosphor bronzes		85	160	180
Cupro-nickels		90		
Nickel silvers		100	125	160
	W TB00	W(8H) TD02	W(H) TD04	W(P) W(½H)P W(H)P TF00 TL02 TL04
Beryllium bronze	110	210	250	380 400 420

Impact properties

At 20°C, annealed tough pitch copper has an Izod value of 47 J, annealed deoxidized copper 61 J. Hard drawn 4% phosphor bronze has an Izod value of 62 J, 7% annealed aluminium bronze 32 J, annealed 80/20 cupro-nickel 104 J. Annealed naval brass has a Charpy value of 82 J, annealed silicon bronze 90 J, annealed 70/30 cupro-nickel 90 J, annealed nickel silver (30% Ni) 108 J.

Machinability

Copper and its alloys can be considered to fall into three groups with regard to machining.

Group 1: Free cutting, with machinability rating greater than 70%.

Group 2 Readily machinable, with machinability rating greater than 30% but less than 70%.

Group 3 Difficult to machine, with machinability rating less than 30%.

On the British Standards system free machining brass (CZ121) is given a rating of 100%, on the Copper Development Association system free cutting brass (C360) is rated as 100%. Machinability ratings according to the above groups are given in the tables relating to mechanical properties. See Mechanical properties of cast alloys and Mechanical properties of wrought alloys.

Mechanical properties of cast alloys

Tables 5.14 and 5.15 show typical properties of commonly used cast copper and its alloys. The tensile modulus of copper and high copper alloys is 120 GPa, for cast brasses 90–100 GPa, cast tin bronzes 70–80 GPa, cast gunmetals 80 GPa, cast aluminium bronzes 100–120 GPa (increasing with increasing alloy content), cast cupro-nickels 120–150 GPa (increasing with increasing nickel) and cast nickel silvers 120–137 GPa (increasing with increasing nickel).

Table 5.14 Properties of cast BS alloys

Name	BS code	IACS (%)	Yield stress (MPa)	Tensile strength (MPa)	Elong-ation (%)	Mach. group
Copper						
High conductivity	HCC1	90	30	155	25	
Brasses						
Sand-cast brass	SCB1	18	80–110	170–200	18–40	1
Sand-cast brass	SCB3	20	70–110	190–220	11–30	1
Sand-cast naval brass	SCB4	18	70–105	250–310	18–40	2
Brazable sand-cast br.	SCB6	25	80–110	170–190	18–40	3
Gravity die-cast br.	DCB1	18	90–120	280–370	23–50	2
Gravity die-cast naval	DCB3	18	90–120	300–340	13–40	1
Pressure die-cast br.	PCB1	18	90–120	280–370	25–40	1
High tensile brass	HTB1	22	170–280	470–570	18–35	2
High tensile brass	HTB3	8	400–470	740–810	11–18	2
Bronzes						
Tin bronze	CT1	9	130–160	230–310	9–20	3
Phosphor bronze	PB1	9	130–230	220–420	3–22	3
Phosphor bronze	PB2	9	130–200	220–370	5–15	3
Phosphor bronze	PB4	9	100–230	190–400	3–20	3
Leaded phosphor bronze	LPB1	11	80–200	190–360	3–18	1
Leaded br. 76-9-0-15	LB1	11	80–190	170–310	4–10	1
Leaded br. 80-10-0-10	LB2	10	80–220	190–390	5–15	1
Leaded br. 85-5-0-10	LB4	13	60–170	160–310	7–20	1
Leaded br. 75-5-0-20	LB5	14	60–160	160–270	5–16	1
Gunmetals						
Admiralty gunmetal	G1	15	130–170	250–340	5–25	2
Nickel gunmetal	G3	12	140–160	280–340	16–25	2
Leaded g.m 83-3-9-5	LG1	16	80–140	180–340	11–35	1
Leaded g.m 85-5-5-5	LG2	15	130–160	250–370	13–25	1
Leaded g.m 87-7-3-3	LG4	13	130–160	250–370	13–30	1
Aluminium bronzes						
Aluminium bronze	AB1	13	170–200	500–590	18–40	3
Aluminium bronze	AB2	8	250–300	640–700	13–20	3
Cu–Mn–Al	CMA1	3	280–340	650–730	18–35	3
Nickel-silvers						
Cu–Ni–Cr	CN1			200–270	10–25	3
Cu–Ni–Nb	CN2			340–450	15–25	3

Note: The yield stress is the 0.1% proof stress, except for ABI where it is the 0.2% proof stress. For an explanation of machinability group, see Machinability.

Table 5.15 Properties of cast CDA alloys

Name	CDA	IACS (%)	Yield stress (MPa)	Tensile strength (MPa)	Elong-ation (%)	Mach. group
Copper						
	C801		62	170	40	3
	C811		62	170	40	3
High copper alloys						
	C817		470	630	8	2
	C821		470	630	8	2
Brasses						
Leaded red brass	C836	15	105	240	32	1
Leaded red brass	C838		85–115	205–260	15–27	1

Leaded semi-red brass	C844	18	90–115	200–270	18–30	1
Leaded semi-red brass	C848	17	105	260	37	1
Leaded yellow brass	C852	15–22	85–95	240–275	25–40	1
Leaded yellow brass	C854	18–25	75–105	205–260	20–35	1
Leaded naval brass	C857	20–26	95–140	275–310	15–25	1
Leaded yellow brass	C858		210	380	15	1
Manganese bronzes						
Manganese bronze	C862	7–8	315–345	625–670	19–25	2
Manganese bronze	C863	9	570	820	18	3
Tin bronzes						
Tin bronze	C903	12–13	125–150	275–345	25–50	2
Tin bronze	C905	10–12	140–160	275–345	24–43	2
Tin bronze	C911		170	240	2	3
Steam bronze/valve bronze	C922	15	110	280	45	2
Leaded tin bronze	C923	10–12	110–165	225–295	18–30	2
High-lead tin bronze	C932		115–145	205–260	12–20	1
High-lead tin bronze	C937	10	125	270	30	1
High-lead tin bronze	C938	95–140	170–225	10–18		1
High-lead tin bronze	C943	75–105	160–205	7–16		1
Nickel-tin bronze	C947	140–34				
		5	310–515	5–25	2	
Aluminium bronzes						
Aluminium bronze	C952	12–14	170–205	480–600	22–38	2
Aluminium bronze	C953	12–15	205–380	480–655	12–35	2
Nickel aluminium bronze	C955	8–10	275–550	620–855	5–20	2
Nickel aluminium bronze	C958		260	655	25	2
Silicon bronzes/brasses						
Silicon brass	C875	6	210	470	17	2
Nickel–silvers						
Nickel–silver	C973	6	105–140	205–275	10–25	1
Nickel–silver	C976	5	180	325	22	1
Nickel–silver	C978	4–5	180–275	345–450	15–25	2

Note: The yield stress is the 0.1% proof stress. For an explanation of machinability group see Machinability.

Mechanical properties of wrought alloys

Tables 5.16 and 5.17 give typical properties of wrought copper alloys. The tensile modulus of copper and high copper alloys is 120 GPa, for brasses, 100–120 GPa (decreasing with increasing zinc content and increasing with increasing cold work), phosphor bronzes, 110 to 120 (decreasing with increasing tin content and decreasing with increasing cold work), aluminium bronzes, 114–140 GPa (decreasing with increasing aluminium content and decreasing with increasing cold work), cupro–nickels, 125–155 GPa (increasing with increasing nickel content and decreasing with increasing cold work), nickel silvers, 120–137 GPa (increasing with increasing nickel content and increasing with increasing cold work) and beryllium bronzes, 120–125 GPa when solution treated and 135–140 GPa when precipitation hardened.

Table 5.16 Properties of wrought alloys

Name	BS	IACS (%)	Yield stress (MPa)	Tensile strength (MPa)	Elong-ation (%)	Mach. group
Copper						
Elec. tough pitch h.c. Cu	C101	101	60–325	220–385	4–55	3
Oxygen free h.c. copper	C103	101	60–325	220–385	4–60	3
Tough pitch arsenical Cu	C105	95–89	60–325	220–385	4–55	

Alloy	Code		Yield stress	Tensile	Elongation	Mach. group
Phosphorus deoxidized Cu	C106	90–70	60–325	220–385	4–60	3
Phosphorus deox. ars. Cu	C107	50–35	60–325	220–385	4–60	3
Oxygen free h.c copper	C110	101	60–325	220–385	4–60	3
High copper alloys						
Copper-cadmium	C108	75–90	60–460	280–700	4–45	3
Brasses						
Gilding metal (90/10 br.)	CZ101	40–44	77–385	265–450	4–50	3
Red brass (85/15 br.)	CZ102	32–37	90–400	290–46	10–60	3
70/30 arsenical brass	CZ105	28	120	330		3
Deep drawing brass (70/30)	CZ106	22–28	110–450	325–54	15–70	3
Basis or common brass	CZ108	22–26	130–180	340–55	5–55	2
Muntz metal (60/40 br.)	CZ109	24–28	160–450	380–500	10–30	1
Aluminium brass	CZ110	23	125–155	340–390	50–60	3
Admiralty brass	CZ111	25	110–420	320–520	10–60	3
Naval brass	CZ112	26	140–340	370–525	20–40	2
High tensile brass	CZ114	21	210	480	20	2
High tensile soldering br.	CZ115	18	210	480	20	1
High tensile brass	CZ116	12	300	570	18	2
Leaded brass (clock br.)	CZ118	26	108–325	325–525	7–50	1
Leaded Muntz metal	CZ123	20–27	160–400	380–520	5–40	2
Phosphor bronzes						
4% Phosphor bronze	PB101	15–25	110–460	320–590	8–55	3
5% Phosphor bronze	PB102	12–18	120–520	340–700	8–60	3
7% Phosphor bronze	PB103	10–15	140–570	370–650	14–65	3
Aluminium bronzes						
7% Aluminium bronze	CA102	15–18	90–230	420–540	10–50	3
9% Aluminium bronze	CA103	12–14	260–340	570–650	22–30	3
10% Aluminium bronze	CA104	7–9	430–600	850–900	15–25	3
Aluminium br.(alloy E)	CA105	7–9	260–400	660–770	17–22	3
Aluminium br.(alloy D)	CA106	4–15	230–270	540–670	22–40	3
Silicon bronze						
Copper silicon	CS101	6–7	75–305	360–540	35–78	2
Beryllium bronzes						
Copper-beryllium(sol.tr.)	CB101	6–78	185–190	480–500	45–50	3
+ precipitate		22–32	930–940	1150–1160	5	
Cupro-nickels						
95/5 Cupro-nickel	CN101	12	115–310	390–430	9–43	3
90/10 Cupro-nickel	CN102	9	120–380	320–420	12–42	3
80/20 Cupro-nickel	CN104	6	120–390	340–450	15–40	3
75/25 Cupro-nickel	CN105	5	140–390	360–450	15–40	3
70/30 Cupro-nickel	CN107	5	150–430	390–500	16–42	3
Special 70/30 Cupro-nick.	CN108	5	170–570	420–660	7–42	3
Nickel silvers						
Leaded nickel brass	NS101	9	180–280	460–590	10–30	1
10% nickel silver	NS103	8	100–600	350–690	5–65	3
12% nickel silver	NS104	8	110–600	350–710	4–60	3
15% nickel silver	NS105	7	130–630	360–710	4–55	3
18% nickel silver	NS106	6	120–630	390–710	5–52	3
18% nickel silver	NS107	6	160–710	390–790	3–48	3
20% leaded nickel brass	NS111	7				

Note: The yield stress is the 0.1% proof stress, except for CZ105, CZ108, CZ109, CZ211 and CB101, where it is the 0.2% proof stress. The spread of values of the properties is determined by the condition of the material. The lower values of the yield and tensile stresses correspond to the soft or annealed state, the higher values to the hard state. The higher values of the elongation correspond to the soft state, the lower values to the hard state. For an explanation of machinability group see Machinability.

Table 5.17 Properties of CDA wrought copper alloys

Name	CDA	IACS (%)	Yield stress (MPa)	Tensile strength (MPa)	Elong- ation (%)	Mach. group
Copper						
Oxygen-free electronic	C101	101	70–365	220–455	4–55	3
Oxygen-free copper	C102	101	70–365	220–455	4–55	3
Electrolytic tough pitch	C110	101	70–365	220–455	4–55	3
Phosphorus deoxidized Cu	C122		70–345	220–380	8–45	3
Phosphorus deox. arsenical	C142		70–345	220–380	8–45	3
Deoxidized cadmium copper	C143		75–385	220–400	1–42	3
High copper alloys						
Cadmium copper	C162		50–475	240–690	1–57	3
Chromium copper	C182		97–530	235–590	4–40	3
Brasses						
Gilding metal	C210		70–400	235–440	4–45	3
Commercial bronze, 90%	C220		70–427	255–500	3–50	3
Red brass	C230		70–435	270–630	3–55	2
Cartridge brass, 70%	C260	28	75–450	300–900	3–66	2
Yellow brass	C268–0	27	100–430	320–880	3–65	2
Muntz metal	C280		145–380	375–510	10–52	2
Free-cutting brass	C360		125–310	340–470	18–53	1
Free-cutting Muntz metal	C370		140–415	370–550	6–40	1
Inhibited admiralty	C443–5		125–150	330–380	60–65	2
Naval brass	C464–7		170–455	380–610	17–50	2
Manganese brass	C674		235–380	480–630	20–28	3
Aluminium brass, arsenical	C687		186	414	55	2
Phosphor bronzes						
Phosphor bronze, 1.25% E	C505	48	100–345	275–545	4–48	3
Phosphor bronze, 5% A	C510		130–550	325–965	2–64	3
Phosphor bronze	C511		345–550	320–710	2–48	3
Phosphor bronze, 8%	C521		165–550	380–965	24–80	3
Phosphor bronze, 10%	C524		193	455–1014	3–70	3
Aluminium bronzes						
Aluminium bronze, 5%	C608		186	414	55	3
Aluminium bronze	C613		210–400	480–585	35–42	2
Aluminium bronze, D	C614		230–415	525–615	32–45	3
Aluminium bronze	C623		240–360	520–675	22–35	2
Aluminium bronze	C630		345–520	620–815	15–20	2
Silicon bronze						
Low silicon bronze B	C651	12	100–475	275–655	11–55	2
High silicon bronze, A	C655		145–485	385–1000	21–70	2
Beryllium bronzes						
Beryllium copper	C172	22–30	170–1345	470–1460	1–48	3
Cupro-nickels						
Copper-nickel, 10%	C706		110–390	300–415	10–42	3
Copper-nickel, 20%	C710		90–585	340–655	3–40	3
Copper-nickel, 30%	C715		140–480	370–520	15–45	3
Copper-nickel	C717		210–1240	480–1380	4–40	3
Copper-nickel	C725		150–745	380–830	1–35	3
Nickel silvers						
Nickel silver, 65–10	C745		125–525	340–895	1–50	3
Nickel silver, 65–18	C752		170–620	385–710	3–45	3
Nickel silver, 65–15	C754		125–545	365–635	2–43	3
Nickel silver, 65–12	C757		125–545	360–640	2–48	3
Nickel silver, 55–18	C770	6	185–620	415–1000	2–40	2
Leaded ni. silver, 65–8–2	C782		160–525	365–630	3–40	2

Note: The yield stress values are for 0.1% proof stress. The range of values correspond to the range of tempers from soft to hard. The low values of yield and tensile strength correspond to the soft or annealed condition, the high values to the hard condition. Conversely, the low values of the elongation correspond to the hard condition and the high values to the soft condition. For an explanation of machinability group, see Machinability.

Solderability

Table 5.18 indicates the relative effectiveness of brazing and soldering of copper and its alloys.

Table 5.18   Brazing and soldering of copper alloys

Alloy	Soldering	Brazing
Copper		
Oxygen free	VG	VG
Deoxidized	VG	VG
Tough pitch	VG	G
Brasses		
Red	VG	VG
Yellow	VG	VG
Leaded	VG	G
Admiralty	VG	VG
Bronzes		
Phosphor	VG	VG
Leaded phosphor	VG	G
Aluminium bronzes		
Aluminium bronzes	P	P
Silicon bronzes		
Silicon bronzes	VG	VG
Beryllium bronzes		
Beryllium copper	G	G
Cupro-nickels		
Cupro-nickels	VG	VG
Nickel silvers		
Nickel silvers	VG	VG

Note: VG = very good, G = good, P = poor.

Thermal properties

Table 5.19 shows typical values of linear thermal expansivity, i.e. coefficient of expansion, and thermal conductivity of copper and its alloys at about 20°C.

Table 5.19   Thermal properties of copper alloys

Material	Thermal conduct. ($W\ m^{-1}\ {}^{\circ}C^{-1}$)	Thermal expansiv. ($10^{-6}\ {}^{\circ}C^{-1}$)
Wrought alloys		
Copper	390	17
High copper alloys	340–370	18
Gilding metal	190	18
Red brass	160	19
Cartridge brass	120	20
Yellow, common, brass	120	21
Muntz metal	125	21
Naval brass	113	21
High tensile brass	88	20

Phosphor bronze, 7%	60	19
Aluminium bronze, 7%	70	17
Aluminium bronze, 9%	42	17
Aluminium bronze, 10%	46	17
Aluminium bronze, alloy D	65	16
Copper-beryllium	120	18
Cupro-nickel, 95/10	67	17
Cupro-nickel, 90/10	50	17
Cupro-nickel, 80/20	38	17
Cupro-nickel, 70/30	29	16
Leaded 10% nickel brass	46	19
10% nickel silver	37	16
12% nickel silver	30	16
15% nickel silver	27	16
Cast alloys		
Copper, high conductivity	370	18
Brass	80-90	19-21
Gunmetal	50	18
Leaded gunmetal	50-80	18
Tin bronze	50	18
Phosphor bronze	47	18-19
Aluminium bronze	42-60	16-17

Weldability

Table 5.20 shows the relative weldability of copper and its alloys.

Table 5.20 Weldability of copper and its alloys

Alloy	Arc welding			Resistance welding			
	Gas met. or tung.	Submer. arc	Shield. metal	Spot	Seam	Proj.	Flash
Coppers							
Oxygen free	G	No	No	No	No	No	G
Deoxidized	VG	No	No	No	No	No	G
Tough pitch	P	No	No	No	No	No	G
Brasses							
Red brass	G	No	No	P	No	P	G
Yellow brass	P	No	No	G	No	G	G
Leaded brass	No	No	No	No	No	No	P
Admiralty brass	G	No	P	G	P	G	G
Bronzes							
Phosphor	G	No	P	G	P	G	VG
Leaded phosphor	No	No	No	No	No	No	P
Aluminium bronzes							
Aluminium bronze	VG	No	G	G	G	G	G
Silicon bronzes							
Silicon bronze	VG	No	P	VG	VG	VG	VG
Cupro-nickels							
10% cupro-nickel	VG	No	G	G	G	G	VG
30% cupro-nickel	P	No	P	VG	VG	VG	VG
Nickel silvers	P	No	No	G	P	G	G

Note: VG = very good, G = good, P = poor, restricted use, No = not to be used.

5.5 Uses

Forms

Tables 5.21 and 5.22 show the forms of wrought copper and alloys that are normally available.

Table 5.21 Forms of BS copper and alloys

Name	BS	Bar	Sheet	Plate	Tube	Wire	Rod
Copper							
Elec. tough pitch h.c. Cu	C101	*	*	*	*	*	*
Oxygen free h.c. copper	C103	*	*	*	*	*	
Tough pitch arsenical Cu	C105		*	*			
Phosphorus deoxidized Cu	C106	*	*	*	*	*	*
Phosphorus deox. ars. Cu	C107		*	*	*		
Oxygen free h.c. copper	C110	*					*
High copper alloys							
Copper-cadmium	C108	*	*	*		*	
Brasses							
Gilding metal (90/10 br.)	CZ101		*	*	*	*	*
Red brass (85/15 br.)	CZ102		*	*	*	*	*
70/30 arsenical brass	CZ105		*	*	*	*	*
Deep drawing brass (70/30)	CZ106		*	*	*	*	*
Basis or common brass	CZ108		*	*	*	*	*
Muntz metal (60/40 br.)	CZ109		*	*	*	*	*
Aluminium brass	CZ110		*	*	*		
Admiralty brass	CZ111				*		
Naval brass	CZ112	*	*	*	*		*
High tensile brass	CZ114						*
High tensile soldering br.	CZ115						*
High tensile brass	CZ116						*
Leaded brass (clock br.)	CZ118	*	*				*
Leaded Muntz metal	CZ123	*	*				*
Phosphor bronzes							
4% Phosphor bronze	PB101	*	*			*	
5% Phosphor bronze	PB102	*	*	*		*	
7% Phosphor bronze	PB103	*	*			*	*
Aluminium bronzes							
7% Aluminium bronze	CA102			*	*		
9% Aluminium bronze	CA103	*					
10% Aluminium bronze	CA104	*					
Aluminium br. (alloy E)	CA105	*	*	*			
Aluminium br. (alloy D)	CA106	*	*				
Silicon bronze							
Copper silicon	CS101	*	*	*	*	*	*
Beryllium bronzes							
Copper-beryllium	CB101	*	*	*	*	*	*
Cupro-nickels							
95/5 Cupro-nickel	CN101	*	*	*			
90/10 Cupro-nickel	CN102	*	*	*			
80/20 Cupro-nickel	CN104	*	*	*			
75/25 Cupro-nickel	CN105	*					
70/30 Cupro-nickel	CN107	*	*	*	*		
Special 70/30 Cupro-nick.	CN108			*			
Nickel silvers							
Leaded nickel brass	NS101						*
10% nickel silver	NS103		*	*		*	

Name	CDA	Flat	Rod	Wire	Tube	Pipe	Shape
12% nickel silver	NS104		*	*		*	
15% nickel silver	NS105		*	*		*	
18% nickel silver	NS106		*	*		*	
18% nickel silver	NS107		*	*		*	
20% leaded nickel brass	NS111	*					*

Table 5.22 Forms of CDA copper and alloys

Name	CDA	Flat	Rod	Wire	Tube	Pipe	Shape
Copper							
Oxygen free electronic	C101	*	*	*	*	*	*
Oxygen free copper	C102	*	*	*	*	*	*
Electrolytic tough pitch	C110	*	*	*	*	*	*
Phosphorus deoxidized Cu	C122	*			*	*	
Phosphorus deox. arsenical	C142	*			*	*	
Deoxidized cadmium copper	C143	*					
High copper alloys							
Cadmium copper	C162	*	*	*			
Chromium copper	C182	*	*	*			*
Brasses							
Gilding metal	C210	*		*			
Commercial bronze, 90%	C220	*	*	*	*		*
Red brass	C230	*		*	*	*	*
Cartridge brass, 70%	C260	*	*	*			
Yellow brass	C268-0	*	*	*			
Muntz metal	C280	*	*				
Free-cutting brass	C360	*	*				*
Free-cutting Muntz metal	C370				*		
Inhibited admiralty	C443-5	*			*	*	
Naval brass	C464-7	*	*		*		*
Manganese brass	C674	*	*				
Aluminium brass, arsenical	C687				*		
Phosphor bronzes							
Phosphor bronze, 1.25% E	C505	*		*			
Phosphor bronze, 5% A	C510	*	*	*	*		
Phosphor bronze	C511	*					
Phosphor bronze, 8%	C521	*	*	*			
Phosphor bronze, 10%	C524	*	*	*			
Aluminium bronzes							
Aluminium bronze, 5%	C608				*		
Aluminium bronze	C613	*	*		*	*	*
Aluminium bronze, D	C614	*	*	*	*	*	*
Aluminium bronze	C623	*	*				*
Aluminium bronze	C630	*	*				
Silicon bronze							
Low silicon bronze B	C651		*	*	*		
High silicon bronze, A	C655	*	*	*	*		
Beryllium bronzes							
Beryllium copper	C172	*	*	*	*	*	*
Cupro-nickels							
Copper-nickel, 10%	C706	*				*	
Copper-nickel, 20%	C710	*			*		
Copper-nickel, 30%	C715	*	*		*		

Copper-nickel	C717	*	*	*	
Copper-nickel	C725	*	*	*	*
Nickel silvers					
Nickel silver, 65-10	C745	*		*	
Nickel silver, 65-18	C752	*	*	*	
Nickel silver, 65-15	C754	*			
Nickel silver, 65-12	C757	*		*	
Nickel silver, 55-18	C770	*	*	*	
Leaded ni. silver, 65-8-2	C782	*			

Uses of cast alloys

Table 5.23 gives typical uses of BS and CDA cast copper and alloys.

Table 5.23   Uses of BS and CDA cast copper alloys

Name	BS CDA	Cast. form	Uses
Copper			
High conductivity	HCC1	S,C	Electrical components.
	C801	S,C,T	As above.
	C811	S,C,T	As above.
High copper alloys			
	C817	S,C,T	Electrical components, stronger and harder than coppers.
	C821	S,C,T	As above.
Brasses			
Leaded red brass	C836	S,C,T	Plumbing items, valves, flanges, pump parts.
Leaded red brass	C838	S,C,T	Plumbing items, low-pressure valves, pump parts.
Leaded semi-red brass	C844	S,C,T	General hardware, plumbing items, low-pressure valves.
Leaded semi-red brass	C848	S,C	As above.
Sand cast brass	SCB1	S,D	Hardware, pressure tight valves and fittings.
Leaded yellow brass	C852	C,T	Plumbing fittings, hardware, valves, ornamental items.
Sand cast brass	SCB3	S,D	Plumbing, gas fittings, electrical components.
Leaded yellow brass	C854	S,C,T	General purpose hardware.
Leaded naval brass	C857	S,C	Hardware, ornamental items.
Sand cast naval brass	SCB4	S	Marine fittings, pumps, heat exchangers.
Brazable sand cast br.	SCB6	S,D	Items to be brazed, marine fittings.
Gravity die cast br.	DCB1	GD,C	Accurate castings for cars, plumbing fittings, etc.
Gravity die cast naval	DCB3	GD,C	Accurate castings, good machinability.
Leaded yellow brass	C858	D	As above.
Pressure die cast br.	PCB1	D	As above.
Manganese bronzes			

High tensile brass	HTB1	S,GD,C	Components subject to high stress at normal temps., e.g. marine propellers, rudders, pumps.
High tensile brass	HTB3	S,C	Components subject to high stress but not stress corrosion. Marine castings.
Manganese bronze	C862	S,D,C,T	Marine castings, bearings.
Manganese bronze	C863	S,C	As HTB3.

Bronzes

Tin bronze	C903	S,C,T	Bearings, gears, piston rings, valve components.
Tin bronze	C905	S,C,T	As above.
Tin bronze	C911	S	Bearings, piston rings.
Tin bronze	CT1	S,D,C,T	General purpose sand castings.
Phosphor bronze	PB1	S,D,C,T	Bearings, gears, heavy duty bushes.
Phosphor bronze	PB2	S,D,C,T	Heavy duty bearings, gears subject to shock loads.
Phosphor bronze	PB4	S,D,C,T	Widely used but lower loads than PB1.
Leaded phosphor bronze	LPB1	S,D,C,T	Lighter duty than PB1,2 or 4, but more able to withstand poor lubrication.
Steam bronze/valve br.	C922	S,C,T	Pressure containing parts up to 300°C.
Leaded tin bronze	C923	S,C,T	Valves, pipes and high pressure steam castings.
High-lead tin bronze	C932	S,C,T	Bearings and bushes.
Leaded br. 76-9-0-15	LB1	D,C,T	Bearings and bushes resisting poor lubrication and corrosion.
Leaded br. 80-10-0-10	LB2	S,D,C,T	As above.
Leaded br. 85-5-0-10	LB4	S,D,C,T	As above.
High-lead tin bronze	C937	S,C,T	High speed and heavy pressure bearings, corrosion resistant items.
High-lead tin bronze	C938	S,C,T	General service and moderate pressure bearings.
High-lead tin bronze	C943	S,C	High speed bearings for light loads.
Leaded br. 75-5-0-20	LB5	T	Steel backed bearings for engines.
Nickel-tin bronze	C947	S,C,T	Bearings, wear guides, piston cylinders, nozzles.

Gunmetals

Admiralty gunmetal	G1	S,D,C	General purpose, e.g. pumps, valves, bearings.
Nickel gunmetal	G3	S,D,C	Bearing cages, switch parts, valves, slow moving parts.
Leaded g.m 83-3-9-5	LG1	S,D,C,T	General purpose where pressure tightness required, e.g. pumps, valves.

Leaded g.m 85-5-5-5	LG2	S,D,C,T	As LG1, but better pressure tightness.
Leaded g.m 87-7-3-3	LG4	S,D,C,T	As LG1, but better strength.
Aluminium bronzes			
Aluminium bronze	AB1	S,GD,C	High strength, high corrosion resistance items, e.g. pumps, marine items. Die cast items for cars.
Aluminium bronze	C952	S,C,T	Bearings, gears, bushes, valve seats, acid-resisting pumps.
Aluminium bronze	C953	S,C,T	Gears, marine equipment, nuts, pickling baskets.
Aluminium bronze	AB2	S,GD,C,T	Higher temp. use than AB1.
Cu-Mn-Al	CMA1	S,GD,C,T	Heavy, intricate items with high resistance to corrosion and wear.
Nickel aluminium br.	C955	S,C,T	Corrosion resistant items, agitators, gears, aircraft valve guides and seats.
Nickel aluminium br.	C958	S,C,T	Corrosion resistant items.
Silicon bronzes/brasses			
Silicon brass	C875	S,C,D	Bearings, gears, rocker arms, impellers.
Nickel-silvers			
Cu-Ni-Cr	CN1	S	Ornamental and hardware items.
Cu-Ni-Nb	CN2	S	As above.
Nickel-silver	C973	S	As above.
Nickel-silver	C976	S	Marine castings and ornamental items.
Nickel-silver	C978	S	Valves and valve seats, musical instrument components, ornamental items.

Note: S = sand cast, C = centrifugal cast, D = die cast, GD = gravity die cast, T = continuously cast.

Uses of wrought alloys

Table 5.24 gives typical uses of BS and CDA wrought copper and alloys.

Table 5.24 Uses of wrought copper and alloys

Name	BS	CDA	Uses
Copper			
Elec. tough pitch h.c. Cu	C101	C110	C/H working. Electrical wires, busbars, heat exchangers, roofing.
Oxygen free h.c. copper	C103	C102	C/H working. Electronic components, waveguides, busbars.
Tough pitch arsenical Cu	C105		C/H working. General engineering work.
Phosphorus deoxidized Cu	C106	C122	C/H working. Plumbing piping, chemical plant, heat exchangers.
Phosphorus deox. ars. Cu	C107	C142	C/H working. Heat exchanger and

			condenser tubes.
Oxygen free h.c. copper	C110	C101	C/H working. High electrical conductivity items for electronics.
Deoxidized cadmium copper		C143	C/H working. Electrical uses requiring thermal softening resistance, contacts, terminals, welded items.

High copper alloys

Copper–cadmium	C108	C162	C/H working. High strength electrical transmission lines, electric blanket elements.
Chromium copper	A2/1	C182	C/H working. Spot welding electrodes, switch contacts, cable connectors.

Brasses

Gilding metal (90/10 br.) or commercial bronze, 90%	CZ101	C220	C/H working. Decorative and architectural items, screws, rivets.
Gilding metal (95/5 br.)	CZ125	C210	C/H working. Coins, medals, ammunition caps.
Red brass (85/15 br.)	CZ102	C230	C/H working. Plumbing pipes, heat exchanger tubes, fasteners, conduit.
70/30 arsenical brass	CZ105		C working. Deep drawn and cold headed items, hardware.
Deep drawing brass (70/30)	CZ106	C260	C working. Deep drawn and cold headed items, locks, hinges, pins, rivets, hardware.
Basis/common/yellow brass	CZ108	C268-0	C/H working. Spinning, presswork, cold heading. Hardware, fasteners, electrical items.
Muntz metal (60/40 br.)	CZ109	C280	H working. Hot forging, pressing, heading and upsetting. Condenser plates, heat exchanger tubes, brazing rod.
Aluminium brass	CZ110	C687	C/H working. Heat exchanger and condenser tubes.
Admiralty brass	CZ111	C443-5	As above.
Naval brass	CZ112	C464-7	H working. Hot forging, pressing, drawing. Marine hardware, rivets, propeller shafts, bolts, welding rod.
High tensile brass	CZ114		H working. Pump parts, architectural sections,

		pressure tubes.
High tensile soldering br.	CZ115	As above.
High tensile brass	CZ116	As above.
Leaded brass (clock br.)	CZ118	C working. Headed, blanked parts, clock and instrument parts.
Free-cutting brass	CZ121 C360	H working. Excellent machinability. Gears, fasteners, screw machine parts.
Leaded Muntz metal	CZ123 C370	H working. Extrusions, hot forging, good machinability. Condenser tube parts.
Manganese brass	C674	H working. Hot forging and pressing. Gears, shafts, wear plates.

Phosphor bronzes

4% Phosphor bronze	PB101	C working. Screws, bolts, rivets, washers, springs and clips, bellows and diaphragms.
5% Phosphor bronze	PB102 C510	C working. Blanking, drawing, heading, stamping. Bellows, springs and clips, cotter pins, diaphragms.
7% Phosphor bronze	PB103	C working. Spring, clips, washers, welding rods and electrodes, pump parts.
Phosphor bronze, 1.25% E	C505	C/H working. Blanking, heading, stamping. Electrical contacts, flexible hose.
Phosphor bronze	C511	C working. Springs, fuse clips, switch parts, welding rod.
Phosphor bronze, 8%	C521	C working. Blanking, drawing, stamping. More severe conditions than 5% phosphor bronze.
Phosphor bronze, 10%	C524	C working. Blanking, forming, bending. Heavy duty items, e.g. springs, washers, clutch plates.

Aluminium bronzes

7% Aluminium bronze	CA102 C613	C working. Nuts, bolts, corrosion resistant vessels, marine parts.
9% Aluminium bronze	CA103 C623	H working. Hot forging and pressing. Valve bodies and spindles, bolts, bearings, worm gears, cams.
10% Aluminium bronze	CA104 C630	H working. Hot forging and forming. Nuts,

		bolts, valve seats and guides, pump shafts.
Aluminium br.(alloy E)	CA105	As above.
Aluminium br.(alloy D)	CA106 C614	As above.
Aluminium bronze, 5%	CA101 C608	C working. Condenser and heat exchanger tubes.

Silicon bronze

Copper silicon or low-silicon bronze B	CS101 C651	C/H working. Blanking, or drawing, heading, hot forging and pressing. Chemical plant, marine hardware, bolts, cable clamps, hot water tanks and fittings.
High silicon bronze, A	C655	As above.

Beryllium bronzes

Copper-beryllium	CB101 C172	C/H working. High strength, high conductivity items. Springs, clips, fasteners, non-sparking tools.

Cupro-nickels

95/5 Cupro-nickel	CN101	C/H working. Sea-water corrosion resistant items, ammunition.
90/10 Cupro-nickel	CN102 C706	As above.
80/20 Cupro-nickel	CN104 C710	C/H working. Deep drawn and pressing items.
75/25 Cupro-nickel	CN105	C/H working. Coins and medals.
70/30 Cupro-nickel	CN107 C715	C/H working. Salt water piping, fasteners.
Special 70/30 Cupro-nick.	CN108	C/H working. Very good resistance to corrosion erosion in sea water. Mooring cables, ocean cable pins.
Copper-nickel	C717	As above.
Copper-nickel	C725	C/H working. Blanking, drawing, forming, heading, spinning, stamping. Relay and switch springs, bellows

Nickel silvers

Leaded nickel brass	NS101	H working. Clock and watch components, architectural items.
10% nickel silver	NS103 C745	C working. Blanking, drawing, forming, heading, spinning. Rivets, screws, clips, slide fasteners, decorative items.
12% nickel silver	NS104 C757	As above.
15% nickel silver	NS105 C754	As above.
18% nickel silver	NS106 C752	As above.
18% nickel silver	NS107 C770	C working. Blanking, forming. Optical goods,

		springs and clips for electrical equipment.
20% leaded nickel brass	NS111C782	C working. Machined items to resist wear and corrosion, screws, hinges, keys, watch parts.

Note: C = cold, H = hot in relation to good working characteristics.

6 Magnesium

6.1 Materials

Magnesium

Magnesium has a density at 20°C of 1.7×10^3 kg m^{-3} and thus a low density compared with other metals. It has an electrical conductivity of about 60% of that of copper, as well as a high thermal conductivity. It has a tensile strength which is too low for engineering purposes and so is used only in engineering in alloy form. Under ordinary atmospheric conditions magnesium has good corrosion resistance, as a result of an oxide layer which develops on its surface in air. However, this layer is not completely impervious, particularly in air that contains salts, and thus the corrosion resistance can be low under adverse conditions.

Magnesium alloys

Because of the low density of magnesium, the magnesium–base alloys have low densities. They thus find a use in applications where lightness is a prime consideration, e.g. in aircraft. Aluminium alloys have higher densities than magnesium alloys but can have greater strength. The strength-to-weight ratio for magnesium alloys is, however, greater than that of aluminium alloys. Magnesium alloys also have the advantage of good machinability and weld readily. The corrosion resistance is, however, not as good as that of aluminium alloys.

There are three main groups of magnesium alloys in common use, some being mainly used as casting alloys, some as wrought alloys and some offering alloys which can be used in both ways:

1 Magnesium–manganese alloys.
   These are mainly used for sheet metal fabrication processes, being readily welded.
2 Magnesium–aluminium–zinc alloys.
   These are mainly used for sand casting, gravity die casting, extrusions and forging. Solution treatment and precipitation hardening are possible.
3 Magnesium–zirconium–thorium/zinc/silver alloys.
   The zirconium has an intense grain refining effect. The alloys are used in both cast and wrought forms. They have high proof stresses, good resistance to impact, good fabrication properties and good corrosion resistance.

See Codes for composition, Coding system for temper, Composition of cast alloys, Composition of wrought alloys, Annealing, Solution treatment and aging, Stress relieving, Density, Electrical properties, Fatigue properties, Mechanical properties of cast alloys, Mechanical properties of wrought alloys, Thermal properties, Weldability, Forms, Uses of cast alloys, Uses of wrought alloys.

6.2 Codes and compositions

Codes for composition

A widely used coding system is that of the American Society for Testing Materials (ASTM). Letters and numbers are used to designate an alloy. Firstly two letters are used to indicate the major alloying elements (see Table 6.1). These are followed by two numbers to indicate the nominal percentage amounts of these two alloying elements. The third part of the designation is a letter from A to Z, but excepting I and O, to indicate different alloys with the same nominal composition specified by the preceding letters and numbers. The final part of the coding is a specification of the temper, this being the same system as for aluminium. For details of commonly used tempers with magnesium, see Codes for temper.

To illustrate the above coding system, consider the magnesium alloy AZ81A–T4. The principal alloying elements are aluminium (A) and zinc (Z). There is nominally 8% aluminium and 1% zinc. The specific alloy in this type is alloy A. It has a temper of T4, i.e. it is in the solution heat treated condition.

Table 6.1 ASTM code letters for magnesium alloys

Letter	Element
A	Aluminium
B	Bismuth
C	Copper
D	Cadmium
E	Rare earth
F	Iron
H	Thorium
K	Zirconium
L	Beryllium
M	Manganese
N	Nickel
P	Lead
Q	Arsenic
R	Chromium
S	Silicon
T	Tin
Z	Zinc

The coding system used by British Standards uses a different system for cast and wrought magnesium alloys. For cast alloys the designation consists of the letters MAG followed by a number, the number being used to indicate the casting alloy concerned. MAG 3 is an example of a casting alloy. This is sometimes followed by a letter to indicate the temper, the same letters being used as for the British Standards' system for aluminium (see Coding system for temper). For wrought alloys the designation consists of the letters MAG followed by a dash and a letter to indicate the form of supply (S = plate, sheet and strip, E = bars, sections and tubes, including extruded forged stock). Following another dash there is a number upwards from 100 to indicate the specific alloy concerned. Finally there is a letter to indicate the temper, the same letters being used as for the British Standards' system for aluminium (see Coding system for temper).

As an example of the British Standards' system consider the magnesium alloy MAG-S-101M. The MAG indicates that it is a magnesium based alloy. The S indicates that it is in plate, sheet or strip form. The number 101 indicates the specific alloy concerned. The final M indicates that the temper is – as manufactured. An example of a cast alloy is MAG 3 TB, the 3 indicating the specific alloy concerned and the TB the temper, it being solution treated.

Coding system for temper

The coding systems used for temper are the same as those used for aluminium, the British Standards' system being a slight variation from the widely used American CDA system. Table 6.2 gives the tempers generally used with magnesium.

Table 6.2 Coding system for temper

American	British	Temper
F	M	As fabricated/manufactured.
O	O	Annealed.
H10, H11		Slightly strain hardened.
H23, H24, H26		Strain hardened and partially annealed.
T4	TB	Solution heat treated.
T5	TE	Precipitation treated, i.e. artificially aged.
T6	TF	Solution heat treated and precipitation treated, i.e. artificially aged.

Composition of cast alloys

Table 6.3 shows the composition of cast magnesium alloys to ASTM and British Standard specifications.

Table 6.3 Composition of cast magnesium alloys

BS code	ASTM code	Cast. proc.	Mg	Al	Zn	Mn	Th	Zr	Others
Magnesium-aluminium-zinc alloys									
	AZ63A	S,P	rem	6.0	3.0	0.2			
MAG 1	AZ81A	S,P	rem	7.6	0.7	0.2			
MAG 2		S,P	rem	8.2	0.7	0.5			
MAG 3	AZ91C	S,P	rem	9.0	0.7	0.3			
MAG 3	AZ91A	D	rem	9.0	0.7	0.1			
	AZ92A	S,P	rem	9.0	2.0	0.1			
MAG 7		S,P	rem	8.0	0.9	0.5			
Magnesium-aluminium-manganese alloys									
	AM100A	S,P	rem	10.0		0. 1			
	AM60A	D	rem	6.0		0. 1			
Magnesium-zinc-zirconium-rare earth/thorium alloys									
MAG 4	ZK51A	S,P	rem		4.6			0.7	
	ZK61A	S,P	rem		6.0			0.7	
MAG 5	ZE41A	S,P	rem		4.2			0.7	1.2 RE
	ZE63A	S,P	rem		5.8			0.7	2.6 RE
MAG 6	EZ33A	S,P	rem		2.7			0.7	3.3 RE
MAG 8	HZ32A	S,P	rem		2.1		3.3	0.7	
MAG 9	ZH62A	S,P	rem		5.7		1.8	0.7	
Magnesium-zirconium-thorium alloys									
	HK31A		rem				3.3	0.7	
Magnesium-zirconium-silver alloys									
	QE22A	S,P	rem					0.7	2.5 Ag, 2.1 RE

Note: rem = remainder, RE = rare earths.

Composition of wrought alloys

Table 6.4 shows the composition of commonly used wrought magnesium alloys.

Table 6.4 Composition of wrought magnesium alloys

BS code	ASTM code	Nominal composition (%)					
		Mg	Al	Zn	Mn	Th	Zr
Magnesium-aluminium-manganese alloys							
MAG-101	M1A	rem			1.2		
Magnesium-aluminium-zinc alloys							
	AZ10A	rem	1.2	0.4	0.2		
MAG-111	AZ31B	rem	3.0	1.0	0.2		
MAG-121	AZ61A	rem	6.0	1.0	0.2		
Magnesium-zinc-manganese alloys							
MAG-131		rem		1.9	1.0		
Magnesium-zinc-zirconium alloys							
MAG-141		rem		1.0			0.6
	ZK21A	rem		2.3			0.5
MAG-151		rem		3.3			0.6
	ZK40A	rem		4.0			0.5
MAG-161	ZK60A	rem		5.5			0.6
Magnesium-thorium-zirconium/manganese alloys							
	HK31A	rem				3.0	0.6
	HM21A	rem			0.6	2.0	
	HM31A	rem			1.2	3.0	

Note: rem = remainder.

6.3 Heat treatment

Annealing

Annealing of the wrought magnesium-aluminium-zirconium alloys is 1 hour or more at about 345°C, the wrought magnesium-thorium-zirconium/manganese alloys 400 to 450°C, and the wrought magnesium-zinc-zirconium alloys about 290°C.

Solution treatment and aging

Table 6.5 outlines the types of heat treatment used with casting alloys.

Table 6.5 Solution treatment and aging of cast magnesium alloys

	Alloy		Solution tr.		Aging		Temper	
Type	Example		temp.°C	time h	temp.°C	time h	BS	USA
Mg-Al-Mn	AM100A				230	5	TE	T5
Mg-Al-Zn	AZ91C				170	16	TE	T5
	AZ91C,	MAG 3	415	16–20*			TB	T4
	AZ81A,	MAG 1,						
		MAG 7.						
	AZ91C,	MAG 3,	415	16–20*	170	16	TF	T6
		MAG 7.						
Mg-Zn-X	EZ33A				215	5	TE	T5
		MAG 4			175	10–16	TE	T5
		MAG 6						
	HZ32A,	MAG 8			315	16	TE	T5
	ZE41A,	MAG 5			330	2 and		
					175	16	TE	T5
	ZH62A,	MAG 9			330	2 and		
					175	16	TE	T5

Note: * indicates alternative forms of treatment are possible.

Stress relieving

Stress relief of the wrought magnesium–aluminium–zirconium alloys is about 1 hour at 150 to 200°C for hard rolled sheet or 15 minutes at 200 to 260°C for extrusions and forgings. For the wrought magnesium–thorium–zirconium/manganese alloys it is 30 minutes at 300 to 400°C. For the wrought magnesium–zinc–zirconium alloys it is 150 to 260°C for 15 minutes to 1 hour.

6.4 Properties

Density

Cast magnesium alloys have densities between 1800 and 1817 kg m^{-3} at 20°C, wrought metal alloys being about 1760 to 1800 kg m^{-3}.

Electrical properties

The electrical conductivity of cast magnesium–aluminium–zinc alloys at 20°C is about 7×10^6 Ω^{-1} m^{-1}, and that of cast magnesium–zirconium etc. alloys about 14 to 15×10^6 Ω^{-1} m^{-1}. For wrought alloys, the magnesium–aluminium–manganese alloys have conductivities of about 20×10^6 Ω^{-1} m^{-1}, the magnesium–aluminium–zinc about 10×10^6 Ω^{-1} m^{-1}, and the magnesium–zinc–zirconium about 16 to 19×10^6 Ω^{-1} m^{-1}.

Fatigue properties

For cast magnesium alloys, the endurance limit at 50×10^6 cycles with cyclic rotating bending loads is about 70 to 100 MPa when unnotched and 50 to 90 MPa when notched.

Mechanical properties of cast alloys

Table 6.6 shows the mechanical properties of cast magnesium alloys at 20°C. The tensile modulus of magnesium alloys tends to be about 40 GPa.

Table 6.6 Mechanical properties of cast magnesium alloys

BS code	ASTM code	Condition BS	Condition ASTM	Tensile strength (MPa)	Yield Tens. (MPa)	Yield Comp (MPa)	Elongation (%)	Hardness (HB)
Magnesium–aluminium–zinc alloys								
	AZ63A	TF	T6	275	130	130	5	73
MAG1	AZ81A	TB	T4	275	83	83	15	55
MAG2		TB	T4	275	83	83	15	55
MAG3	AZ91C	TF	T6	275	195	145	6	66
MAG3	AZ91A	M	F	230	150	165	3	63
	AZ92A	TF	T6	275	150	150	3	84
MAG7		TB	T4	260	83	83	5	55
Magnesium–aluminium–manganese alloys								
	AM100A	TF	T6	275	150	150	1	70
	AM60A	M	F	205	115	115	6	
Magnesium–zinc–zirconium–rare earth/thorium alloys								
MAG4	ZK51A	TE	T5	205	165	165	4	65
	ZK61A	TE	T5	310	185	185		68
MAG5	ZE41A	TE	T5	205	140	140	4	62
	ZE63A	TF	T6	300	190	195	10	70
MAG6	EZ33A	TE	T5	160	110	110	3	50
MAG8	HZ32A	TE	T5	185	90	90	5	55
MAG9	ZH62A	TE	T5	260	170	170	5	70

	Magnesium–zirconium–thorium alloys							
HK31A	TF	T6	220	105	105	8	55
	Magnesium–zirconium–silver alloys							
QE22A	TF	T6	260	195	195	3	80

Note: The yield stress values are the 0.2% proof stress and are given for both tension and compression.

Mechanical properties of wrought alloys

Table 6.7 shows the mechanical properties of wrought magnesium alloys at 20°C. The values given should only be taken as indicative of the order of magnitude, since the values depend on the form of the product, i.e. sheet, bar, tube, and the thickness. The tensile modulus of magnesium alloys tends to be about 40 GPa.

Table 6.7 Mechanical properties of wrought magnesium alloys

BS code	ASTM code	Condition BS	Condition ASTM	Tensile strength (MPa)	Yield Tens. (MPa)	Yield Comp. (MPa)	Elongation (%)	Hardness (HB)
Magnesium–aluminium–manganese alloys								
MAG–101	M1A	M	F	255	180	83	12	40
Magnesium–aluminium–zinc alloys								
	AZ10A	M	F	240	145	70	10	
MAG–111	AZ31B	M	F	260	200	97	15	50
MAG–121	AZ61A	M	F	310	230	130	12	60
Magnesium–zinc–manganese alloys								
MAG–131		M	F	260	200		)8	
Magnesium–zinc–zirconium alloys								
MAG–141		M	F	270	205		)8	
	ZK21A	M	F	260	195	135	4	
MAG–151		M	F	310	225		)8	
	ZK40A	TE	T5	275	255	140	4	
MAG–161	ZK60A	TE	T5	365	305	250	11	90
Magnesium–thorium–zirconium/manganese alloys								
	HK31A		H24	255	200	160	9	70
	HM21A		T8	235	170	130	11	
	HM31A		H24	290	220	180	15	70

Note: The yield stress values given are the 0.2% proof stress.

Thermal properties

The linear thermal expansivity, i.e. coefficient of linear expansion, for both cast and wrought alloys is about 27×10^{-6} °C^{-1}. For cast alloys, the thermal conductivity of magnesium–aluminium–zinc alloys is about 84 W m^{-1} °C^{-1} and that of magnesium–zirconium etc. alloys about 100 to 113 W m^{-1} °C^{-1}. For wrought alloys, it is about 140 W m^{-1} °C^{-1} for magnesium–aluminium–manganese alloys, about 80 W m^{-1} °C^{-1} for magnesium–aluminium–zinc alloys and about 120 to 130 W m^{-1} °C^{-1} for magnesium–zirconium etc. alloys. The specific heat capacity of both cast and wrought alloys is about 1000 J kg^{-1} °C^{-1}.

Weldability

Argon arc welding is the method most used with magnesium alloys. Gas welding must not be used with zirconium containing alloys and special procedures are necessary with other magnesium alloys. See the tables with Uses of cast alloys and Uses of wrought alloys for details about specific alloys.

6.5 Uses

Forms

The forms of casting that can be used with the magnesium casting alloys is shown in Table 6.3, along with the details of alloy composition. Most of the alloys can only be used for sand or permanent mould casting. Table 6.8 shows the forms available for the wrought magnesium alloys.

Table 6.8 Forms of wrought magnesium alloys

BS code	ASTM code	Sheet plate	Extruded bar and shapes
Magnesium–aluminium–manganese alloys			
MAG-101	M1A		*
Magnesium–aluminium–zinc alloys			
	AZ10A		*
MAG-111	AZ31B	*	*
MAG-121	AZ61A		*
Magnesium–zinc–manganese alloys			
MAG-131		*	*
Magnesium–zinc–zirconium alloys			
MAG-141		*	*
	ZK21A		*
MAG-151		*	*
	ZK40A		*
MAG-161	ZK60A		*
Magnesium–thorium–zirconium/manganese alloys			
	HK31A	*	
	HM21A	*	
	HM31A		*

Uses of cast alloys

Table 6.9 shows typical uses of cast magnesium alloys. MAG 1, MAG 3, MAG 4 and MAG 7 can be considered general purpose alloys with MAG 2, MAG 5, MAG 6, MAG 8 and MAG 9 being used for special applications.

Table 6.9 Uses of cast magnesium alloys

BS code	ASTM code	Typical uses
Magnesium–aluminium–zinc alloys		
MAG 1	AZ81A	A general purpose sand and permanent mould casting alloy, suitable for where ductility and resistance to shock are required with moderately high strength. Used for car wheels, portable electric tools, plastics moulds.
MAG 2		A high purity version of MAG 1. Used for instrument casings, photographic and optical equipment.
MAG 3	AZ91C AZ91A	Less microporosity than MAG 1 and so used for pressure-tight applications. Used for car engine covers and manifolds, power tool and electric motor components.
MAG 7		Similar properties to MAG 3 and is very widely

used. Uses include portable power tool items, chain saw parts, bricklayers' hods, heavy vehicle engine sumps, manifolds and covers, car wheels, camera tripods.

Magnesium–zinc–zirconium–rare earth/thorium alloys

MAG 4	ZK51A	A high proof stress alloy with good ductility and mechanical properties up to 150°C. Not suitable for spidery complex castings or welding.
MAG 5	ZE41A	An improvement on MAG 4 where spidery complex castings are required. Used where high strength and pressure tightness are required. Uses include suspension, chassis, bearings and manifold components for racing cars.
MAG 6	EZ33A	Pressure tight and creep resistant up to 250°C and can be argon arc welded.
MAG 8	HZ32A	An improvement on MAG 6 for use up to 350°C.
MAG 9	ZH62A	Higher strength than MAG 4 and so used for heavy duty applications.

Uses of wrought alloys

Table 6.10 shows typical uses of wrought magnesium alloys.

Table 6.10 Uses of cast magnesium alloys

BS code	ASTM code	Typical uses
Magnesium–aluminium–manganese alloys		
MAG-101	M1A	A general purpose alloy with good corrosion resistance, weldable by gas and argon arc, though having low strength.
Magnesium–aluminium–zinc alloys		
MAG-111	AZ31B	A medium strength alloy with good formability and weldable.
MAG-121	AZ61A	A general purpose alloy, weldable by gas and argon arc.
Magnesium–zinc–manganese alloys		
MAG-131		A medium strength alloy with good formability and weldable by argon arc.
Magnesium–zinc–zirconium alloys		
MAG-141		A high strength sheet and extrusion alloy which is fully weldable.
MAG-151		A high strength sheet, extrusion and forging alloy which is weldable under good conditions.
MAG-161	ZK60A	A high strength extrusion and forging alloy but which is not weldable.

7 Nickel

7.1 Materials

Nickel

Nickel has a density of 8.88×10^3 kg m^{-3} and a melting point of 1455°C. It has good tensile strength and maintains it to quite elevated temperatures. It can be both cold and hot worked, has good machining properties and can be joined by welding, brazing and soldering. It possesses excellent corrosion resistance, hence it is often used as a cladding on steel.

Alloys

Nickel is used as the base metal for a number of alloys with excellent corrosion resistance and strength at high temperatures. The alloys can be considered to fall into three main categories:

1 Nickel-copper alloys
Nickel and copper are completely soluble in each other in both the liquid and solid states, see the equilibrium diagram (Figure 5.6). Those nickel–copper alloys containing about 67% nickel and 33% copper are called Monels.

2 Solid solution nickel–chromium, etc. alloys
Nickel-chromium-iron alloys, often with other alloying elements, form a series of solid solution engineering alloys called Inconels and Incoloys. The trade name Hastelloy is often used with solid solution nickel–chromium-molybdenum-iron alloys.

3 Precipitation hardening nickel–chromium-iron, etc. alloys
Alloys based on just nickel–chromium-iron are only hardened by cold working; however the addition of other elements such as aluminium, beryllium, silicon or titanium enables them to be hardened by precipitation heat treatment. The term superalloys is often used for high temperature, heat resistant alloys that are able to maintain their high strengths, resistance to creep and oxidation resistance at high temperatures. They are complex alloys and can be nickel based, nickel–iron based or cobalt based, and include such elements as chromium, cobalt, molybdenum, aluminium, titanium, etc. They are available in both wrought and cast forms. The earliest nickel–base superalloy was Nimonic 80, a nickel–20% chromium solid solution with 2.25% titanium and 1.0% aluminium providing precipitates.

See Codes, Composition of nickel and alloys, Annealing, Solution treatment and precipitation, Stress relief, Creep properties, Density, Electrical resistivity, Fatigue properties, Mechanical properties of cast alloys, Mechanical properties of wrought alloys, Oxidation limit, Thermal properties, Forms, Uses of nickel alloys.

7.2 Codes and compositions

Codes

In general most nickel alloys are referred to by their trade names. British Standards have a system of specifying alloys by the letters NA followed by a number to indicate the specific alloy concerned. Superalloys are given numbers against the letters HR. The American Society for Testing and Materials (ASTM) and the American Society of Mechanical Engineers (ASME) issue specifications for a number of nickel alloys, using a letter or letters followed by a three digit number. Table 7.1 shows the relationships of alloys specified by these methods to the trade names.

Table 7.1 Trade names for nickel alloys

Trade name	BS	ASTM	ASME
Nickel			
Nickel 200	NA 11	B160	SB160
		B163	SB163
Nickel 201	NA 12	B160	SB160
Monel nickel-copper alloys			
Monel 400	NA 13	B127	SB127
		B163-5	SB163-5
		B564	SB395
		F468	SB564
Monel K-500	NA 18		
Solid solution nickel-chromium-iron alloys			
Inconel 600	NA 14	B163	SB163
		B166-8	SB166-8
Incoloy 800	NA 15	B163	SB163
		B407-9	SB407-9
		B564	SB564
Incoloy 800H	NA 15H	B163	SB163
		B407-9	SB407-9
		B564	SB564
Incoloy 825	NA 16	B163	SB163
		B423-5	SB423-5
Incoloy DS	NA 17		
Nimonic 75	HR5		
	HR203		
	HR403		
	HR504		
Precipitation hardening nickel-chromium-iron alloys			
Nimonic 80A	HR1 NA 20	A637	
	HR201		
	HR401		
	HR601		
Nimonic 90	HR2		
	HR202		
	HR402		
	HR501-3		
Nimonic 105	HR3		
Nimonic 115	HR4		
Inco HX	HR6	B435	
	HR204	B572	
		B619	
		B622	
		B626	
Nimonic 263	HR10		
	HR206		
	HR404		
Nimonic 901	HR53		
Nimonic PE 16	HR55		
	HR207		
Inconel X750	HR505	B637	SB637
Casting alloys			
Nimocast 80	ANC9		
Nimocast 90	ANC10		
Nimocast 713	HC203 VMA6A,B		
Nimocast PD21	VMA2		
Nimocast PK24	HC204 VMA12		

Composition of nickel and alloys

Table 7.2 shows the compositions of some of the commonly used wrought nickel alloys and Table 7.3 some of the cast nickel alloys.

Table 7.2 Compositions of wrought nickel alloys

Trade name	Ni	Cr	Fe	Cu	Co	Mo	Al	Ti	Other
Nickel									
Nickel 200	>99.0								<0.15 C
Nickel 201	>99.0								<0.02 C
Monel nickel–copper alloys									
Monel 400	66.5			31.5					
Monel K-500	66.5			29.5			3.0	0.6	
Solid solution nickel–chromium–iron alloys									
Inconel 600	75.0	15.5	8.0						
Incoloy 800	32.5	21.0	45.7				0.4	0.4	
Incoloy 800H	33.0	21.0	45.7				0.4	0.4	0.07 C
Incoloy 825	42.0	21.0	30.0	2.3		3.0		0.9	
Incoloy DS	38.0	18.0	40.5						1.2 Mn, 2.3 Si
Nimonic 75	75.0	19.5	<5.0	<0.5	<5.0			0.4	
Hastelloy X	49.0	22.0	15.8		<1.5	9.0	2.0		0.15 C, 0.6 W
Precipitation hardening nickel–chromium–iron alloys									
Nimonic 80A	73.0	19.5	<1.5	<0.2	<2.0		1.4	2.3	0.07 C
Nimonic 90	55.5	19.5	<1.5	<0.2	18.0		1.5	2.5	<0.13 C
Nimonic 105	54.0	15.0	<1.0	<0.2	20.0	5.0	4.7	1.2	0.15 C
Nimonic 115	55.0	15.0	<1.0	<0.2	14.8	4.0	5.0	4.0	0.16 C
Nimonic 263	51.0	20.0	<0.7	<0.2	20.0	5.9	0.5	2.2	0.06 C
Nimonic 901	44.0	12.5	32.5	<0.2	<1.0	5.8		3.0	0.04 C, 0.015 B
Nimonic PE16	44.0	16.5	23.0	<0.5	<2.0	3.3	1.6	1.2	0.06 C, 0.02 Zr
Astroloy	56.5	15.0	<0.3		15.0	5.3	4.4	3.5	0.06C, 0.03 B, 0.06 Zr
Inco HX	48.5	21.8	18.5		1.5	9.0			0.10 C, 0.60 W
Inconel X750	73.0	15.5	7.0	<0.3			0.7	2.5	0.04 C, 1.0 Nb
René 41	55.0	19.0	<0.3		11.0	10.0	1.5	3.1	0.09 C, 0.01 B
René 95	61.0	14.0	<0.3		8.0	3.5	3.5	2.5	0.16 C, 0.01 B, 3.5 W, 3.5 Nb, 0.05 Zr
Udimet 500	48.0	19.0	<4.0		19.0	4.0	3.0	3.0	0.08C, 0.005 B
Udimet 700	53.0	15.0	<1.0		18.5	5.0	4.3	3.4	0.07C, 0.03 B
Waspaloy	57.0	19.5	<2.0		13.5	4.3	1.4	3.0	0.07 C, 0.006 B, 0.09 Zr

Table 7.3 Compositions of cast nickel alloys

Trade name   Nominal composition %

Trade name	Ni	Cr	Si	Co	Mo	Al	Ti	C	Othe
Nimocast 80	70.0	20.0	0.6	<2.0		1.2	2.6	0.08	0.6 M
Nimocast 90	57.5	19.5	0.6	16.5		1.3	2.4	0.09	0.6 M
									<2.0 F
Nimocast 713	75	13			4	6			2 Ni
Nimocast PD21	75	5.7			2	6			11 W
Nimocast PK24	61	9.5		15	3	5.5	4.7		1
B-1900	64	8		10	6	6	1	0.1	4 T
MAR-M 200	59	9		10		5	2	0.15	1 F
									12.5 W
									1 N
René 77	58	15		15	4.2	4.3	3.3	0.07	
René 80	60	14		9.5	4	3	5	0.17	4

Note: Not all the above alloys have their compositions given in
full detail, some of the very small alloy additions not being
listed.

7.3 Heat treatment

Annealing

Annealing for nickel, such as Nickel 200, involves temperatures in
the range 815 to 925°C and times of about 10 minutes for each
centimetre thickness of material. Thin sheet would be about 3 to
minutes. Monel nickel–copper alloys are annealed at about 870 to
980°C for similar times. The solid solution nickel–chromium–iron
alloys require temperatures in the range 950 to 1175°C with time
ranging from about 5 to 20 minutes per centimetre thickness. The
precipitation hardening nickel–chromium–iron alloys require about
1010 to 1135°C for 10 to 90 minutes per centimetre.

Solution treatment and precipitation

For those nickel–chromium–iron alloys which can be precipitatio
hardened, see Table 7.2; solution treatment generally consists of
about 1060 to 1150°C for times which depend on the allo
concerned, usually 1/2 to 4 hours but with some requiring 8 hour
or even longer. The cooling procedure depends on the allo
concerned, air cooling being used for many though with som
requiring a rapid quench. The precipitation treatment generall
involves temperatures of the order of 620 to 845°C for times rangin
from 2 to 24 hours. Air cooling then generally follows.

Stress relief

Stress relieving treatments are only recommended for nickel, Mon
and a very small number of other nickel alloys. For nickel, e.g
Nickel 200, the treatment is a temperature of 400 to 500°C, fo
Monel about 500 to 600°C. For all other alloys full annealing shoul
generally be undertaken if a stress relieving treatment is require

7.4 Properties

Creep properties

Nickel alloys are widely used for high temperature applications and creep at such temperatures can become a significant problem, generally being the factor limiting the temperature at which the alloy can be used. Table 7.4 shows rupture stresses at different temperatures for a range of nickel alloys.

Table 7.4 Rupture stresses for nickel alloys

Alloy	Temp. (°C)	Rupture stress MPa at 100h	1000 h
Solid solution nickel–chromium–iron alloys			
Inconel 600	815	55	39
	870	37	24
Incoloy 800	650	220	145
	870	45	33
Nimonic 75	870	23	15
Precipitation hardening nickel–chromium–iron alloys			
Nimonic 80A	540		825
	815	185	115
	1000	30	
Nimonic 90	815	240	155
	870	150	69
	1000	40	
Nimonic 105	815	325	225
	870	210	135
	1000	57	
Nimonic 115	815	425	315
	925	205	130
	1000	100	
Nimonic 263	815	170	105
	870	93	46
Inconel X750	540		827
	870	83	45
Udimet 500	815	44	32
Udimet 700	650		102
	815	58	43
Waspaloy	650	108	88
	815	40	25
Cast alloys			
B–1900	815	73	55
	982	26	15
MAR–M 200	815	76	60
	982	27	19
René 80	1000	95	

Density

The density of nickel alloys at 20°C ranges from about 7.8 to 8.9×10^3 kg m^{-3}. For example, Nickel 200 has a density of 8.89×10^3 kg m^{-3}, Monel 400 8.83×10^3 kg m^{-3}, Incoloy 800 7.95×10^3 kg m^{-3}, Incoloy 825 8.14×10^3 kg m^{-3}, Nimonic 90 8.18×10^3 kg m^{-3}, Nimonic 115 7.85×10^3 kg m^{-3}. Nimocast cast alloys have densities of about 8.2 to 8.6×10^3 kg m^{-3}.

Electrical resistivity

The electrical resistivity of nickel, e.g. nickel 200, is about 8 to 19 $\mu\Omega$ m at 20°C. Nickel-copper alloys have resistivities of about 50 $\mu\Omega$ m and nickel–chromium–iron alloys mainly of the order of 110 to 130 $\mu\Omega$ m.

Fatigue properties

The fatigue limit for nickel alloys tends to be about 0.4 times the tensile strength at the temperature concerned. Thus, for example, Udimet 700 has a tensile strength of about 910 MPa at 800°C and a fatigue limit of 340 MPa.

Mechanical properties of cast alloys

Table 7.5 gives the mechanical properties of cast nickel alloys at elevated temperatures.

Table 7.5 Mechanical properties of cast nickel alloys

Name	Temp. (°C)	Tensile strength (MPa)	Yield stress (MPa)	Elong- ation (%)
Nimocast 80	700	970	680	12
	1000	75	40	82
Nimocast 90	700	540	400	18
	900	154	77	40
Nimocast 713	760	940	745	6
	980	470	305	20
Nimocast PD21	700	810	780	2
	1000	560	390	4
Nimocast PK24	700	965	825	6
	1000	500	380	5
B-1900	870	790	700	4
MAR-M 246	870	860	690	5
René 80	870	620	550	11

Note: The values given under yield stress are the 0.2% proof stress

Mechanical properties of wrought alloys

Table 7.6 gives the mechanical properties of wrought nickel alloys at room and elevated temperatures.

Table 7.6 Mechanical properties of wrought nickel alloys

Name	Cond- ition	Temp. (°C)	Tensile strength (MPa)	Yield stress (MPa)	Elong- ation (%)
Nickel					
Nickel 200	CWA	20	380	105	35
	CW	20	535	380	12
Nickel 201	CWA	20	350	80	35
Monel nickel–copper alloys					
Monel 400	CWA	20	480	195	35
Monel K-500	CWSP	20	900	620	20
Solid solution nickel–chromium–iron alloys					
Inconel 600	CWA,HWA	20	550	240	30
	CW	20	830	620	7
		700	365	175	50

		1000	75	40	60
Incoloy 800	CWA,HWA	20	520	205	30
	HW	20	450	170	30
		700	300	180	70
Incoloy 800H	CWS,HWS	20	450	170	30
		700	300	180	70
Incoloy 825	CWA,HWA	20	590	220	30
Incoloy DS		700	335	210	49
		1000	65	35	75
Nimonic 75		20	750		41
		700	420	200	57
		1000	80	50	58
Hastelloy X		20	960	500	43
		650	710	400	37
		870	255	180	50
Precipitation hardening nickel–chromium–iron alloys					
Nimonic 80A	SP	20	1240	620	24
		650	1000	550	18
		870	400	260	34
Nimonic 90	SP	20	1240	805	23
		650	1030	685	20
		870	430	260	16
Nimonic 105	SP	20	1140	815	12
		650	1080	800	24
		870	605	365	25
Nimonic 115	SP	20	1240	860	25
		650	1120	815	25
		870	825	550	18
Nimonic 263	SP	700	750	460	23
		1000	100	75	68
Nimonic 901	SP	700	910	810	12
		1000	90	75	
Nimonic PE16	SP	700	560	400	27
		1000	75	60	
Astroloy	SP	20	1410	1050	16
		650	1310	965	18
		870	770	690	25
Inco HX	SP	700	550	300	44
		1000	150	80	52
Inconel X750	SP	20	1120	635	24
		650	825	565	9
		870	235	165	47
René 41	SP	20	1420	1060	14
		650	1340	1000	14
		870	620	550	19
René 95	SP	20	1620	1310	15
		650	1460	1220	14
Udimet 500	SP	20	1310	840	32
		650	1210	760	28
		870	640	495	20
Udimet 700	SP	20	1410	965	17
		650	1240	855	16
		870	690	635	27
Waspaloy	SP	20	1280	795	25
		650	1120	690	34
		870	525	515	35

Note: CW = cold worked, CWA = cold worked and annealed,
CWS = cold worked and solution treated, HW = hot worked,
HWA = hot worked and annealed, HWS = hot worked and solution

treated, SP = solution treated and precipitation hardened. The values given under the yield stress are the 0.2% proof stress.

Oxidation limit

For nickel–chromium–iron alloys oxidation sets an upper limit of use for the alloys of about 900 to 1100°C.

Thermal properties

The specific heat capacity of nickel and its alloys tends to be in the range 420 to 545 J kg^{-1} °C^{-1} at 20°C. For most the value tends to be about 460 J kg^{-1} °C^{-1}. The linear thermal expansivity, i.e. coefficient of linear expansion, tends to be about 11 to 15×10^{-6} °C^{-1}. The thermal conductivity of nickel, e.g. Nickel 200, is about 75 to 80 W m^{-1} °C^{-1} at 20°C, nickel–copper alloys 22 W m^{-1} °C^{-1}, and nickel–chromium–iron alloys mainly 11 to 12 W m^{-1} °C^{-1}.

7.5 Uses

Forms

Table 7.7 shows the forms available for some nickel wrought alloys.

Table 7.7 Forms of nickel wrought alloys

Name	Form				
	Sheet Plate	Strip	Tube	Wire	Rod, Bar Forgings
Nickel					
Nickel 200	*	*	*	*	*
Nickel 201	*	*	*	*	*
Monel nickel-copper alloys					
Monel 400	*	*	*	*	*
Monel K-500	*	*	*	*	*
Solid solution nickel-chromium-iron alloys					
Inconel 600	*	*	*	*	*
Incoloy 800	*	*	*	*	*
Incoloy 800 H	*		*		*
Incoloy 825	*	*	*		*
Incoloy DS	*	*	*	*	*
Nimonic 75	*	*	*	*	*
Precipitation hardening nickel-chromium-iron alloys					
Nimonic 80A	*	*	*	*	*
Nimonic 90	*	*	*	*	*
Nimonic 105					*
Nimonic 115					*
Nimonic 263	*	*	*	*	*
Nimonic 901					*
Nimonic PE16	*	*			
Inco HX	*	*	*		*

Uses of nickel alloys

Table 7.8 shows typical uses of wrought and cast nickel alloys.

Table 7.8 Uses of nickel wrought and cast alloys

Name	Uses
Nickel	
Nickel 200	Commercially pure nickel. Used for food processing equipment, electrical and electronic

| | parts, equipment for handling caustic alkalis. |
| Nickel 201 | Similar to 200 but preferred for use above 315°C, e.g. caustic evaporators, combustion boats. |

Monel nickel-copper alloys

| Monel 400 | Used for valves and pumps, marine fixtures and fasteners, heat exchangers, fresh water tanks. |
| Monel K-500 | This is age hardenable with high strength and hardness. Used for pump shafts and impellers, valve trim, springs, oil well drill collars and instruments. |

Solid solution nickel-chromium-iron alloys

Inconel 600	Has high oxidation resistance, hence used for high temperature applications. Furnace muffles, heat exchanger tubing, chemical and food processing equipment.
Incoloy 800	Resists hydrogen, hydrogen sulphide corrosion and chloride ion stress corrosion. Used for hydro-carbon cracker tubes, heater element sheaths.
Incoloy 800H	Similar to Incoloy 800 but with improved high temperature strength.
Incoloy 825	High resistance to oxidizing and reducing acids, and sea water. Used in highly corrosive situations, e.g. phosphoric acid evaporators, pickling plant and chemical process items.
Incoloy DS	General purpose heat resisting alloy. Used for furnace parts, heat treatment equipment.
Nimonic 75	Good strength and resistance to oxidation at high temperatures. Used for sheet metal work in gas turbines, furnace parts, heat treatment equipment.

Precipitation hardening nickel-chromium-iron alloys

Nimonic 80A	Used for gas turbine blades and parts, die casting inserts and cores.
Nimonic 90	Used for gas turbine blades and parts, hot working tools.
Nimonic 105	Used for gas turbine blades, discs and shafts.
Nimonic 115	Used for gas turbine blades.
Nimonic 263	Used for gas turbine rings and sheet metal items in service up to 850°C.
Nimonic PE16	Used for items up to 600°C, gas turbine discs and shafts.
Astroloy	Forgings for use at high temperatures.
Inco HX	Used for parts for gas turbines, furnace and heat resistant equipment.
Inconel X750	Used for gas turbine parts, bolts.
René 41	Used for jet engine blades and parts.
Udimet 500	Used for gas turbine parts, bolts.
Udimet 700	Used for jet engine parts.
Waspaloy	Used for jet engine blades.

Cast alloys

B-1900	Used for jet engine blades.
MAR-M 200	Used for jet engine blades.
René 77	Used for jet engine parts.
René 80	Used for turbine blades.

8 Titanium

8.1 Materials

Titanium

Titanium has a relatively low density, 4.5×10^3 kg m^{-3}, just over half that of steel. It has a relatively low strength when pure, but alloying considerably increases it. It has excellent corrosion resistance, but is an expensive metal.

Titanium can exist in two crystal forms, alpha which is a hexagonal close-packed structure and beta which is body centred cubic. In pure titanium the alpha structure is the stable phase up to 883°C and transforms into the beta structure above this temperature. Commercially pure titanium ranges in purity from 99.5 to 99.0%, the main impurities being iron, carbon, oxygen, nitrogen and hydrogen. The properties of the commercially pure titanium are largely determined by the oxygen content.

See Codes for composition, Composition, Annealing, Solution and precipitation treatment, Stress relief, Creep properties, Density, Electrical resistivity, Fatigue properties, Fracture toughness, Hardness, Impact properties, Machinability, Mechanical properties, Thermal properties, Weldability, Forms, Uses.

Titanium alloys

Titanium alloys can be grouped according to the phases present in their structure. The addition of elements such as aluminium, tin, oxygen or nitrogen results in the enlargement of the alpha phase region on the equilibrium diagram, such elements being referred to as alpha-stabilizing. Other elements such as vanadium, molybdenum, silicon and copper enlarge the beta phase region and are known as beta-stabilizing. There are other elements that are sometimes added to titanium and which are neither alpha or beta stabilizers. Zirconium is such an element, being used to contribute solid solution strengthening.

Titanium alloys are grouped into four categories, each category having distinctive properties:

1 Alpha titanium alloys

These are composed entirely of alpha phase, significant amounts of alpha-stabilizing elements being added to the titanium. Such alloys are strong and maintain their strength at high temperatures. They have good weldability but are difficult to work and non-heat treatable.

2 Near-alpha titanium alloys

These are composed of almost all alpha phase with just a small amount of beta phase dispersed throughout the alpha. This is achieved by adding small amounts, about 1 to 2%, of beta stabilizing elements. Such alloys have improved creep resistance at temperatures of the order of 450–500°C.

3 Alpha-beta titanium alloys

These contain sufficient quantities of beta-stabilizing elements for there to be appreciable amounts of beta phase at room temperature. These alloys can be solution treated, quenched and aged for increased strength.

4 Beta titanium alloys

When sufficiently large amounts of beta-stabilizing elements are added to titanium the resulting structure can be made entirely beta at room temperature after quenching, or in some cases air cooling. Unlike alpha alloys, beta alloys are readily cold worked in the solu tion treated and quenched condition and can be subsequently aged to give very high strengths. In the high strength condition the alloys have low ductility. They also suffer from poor fatigue performance.

See Codes for composition, Composition, Annealing, Solution and precipitation treatment, Stress relief, Creep properties, Density, Electrical resistivity, Fatigue properties, Fracture toughness, Hardness, Impact properties, Machinability, Mechanical properties, Thermal properties, Weldability, Forms, Uses.

8.2 Codes and composition

Codes for composition

Titanium alloys are referred to by their structure, i.e. alpha, near-alpha, alpha-beta and beta. Specific alloys within such groups are often just referred to in terms of their nominal composition, e.g. Ti–5Al–2.5Sn for a 5% aluminium–2.5% tin titanium alloy. One general system that is used in Great Britain is the specification of titanium alloys by the company, IMI Ltd. Table 8.1 shows the system. In America there are the ASTM and AMS specifications. In addition, because of the considerable use made of titanium alloys in military applications, there are military systems of specification.

Table 8.1 IMI codes for titanium alloys

Alloy type	Nominal composition	IMI code
Alpha	commercially pure	110, 115, 125, 130, 155, 160
	Ti–Pd	260, 262
Alpha + compound	Ti–2.5Cu	230
Near-alpha	Ti–11Sn–5Zr–2.25Al–1Mo–0.2Si	679
	Ti–6Al–5Zr–0.5Mo–0.25Si	685
	Ti–5.5Al–3.5Sn–3Zr–1Nb–0.25Mo–0.3Si	829
Alpha-beta	Ti–6Al–4V	318
	Ti–4Al–4Mo–2Sn–0.5Si	550
	Ti–4Al–4Mo–4Sn–0.5Si	551
Beta	Ti–11.5Mo–6Zr–4.5Sn	beta III

Composition

The nominal composition of titanium alloys is used to specify alloys in tables of properties given in this chapter. See Mechanical properties for details of the compositions of commonly used titanium alloys.

8.3 Heat treatment

Annealing

Table 8.2 shows typical annealing temperatures and times for titanium alloys. For some alloys, e.g. Ti–8Al–1Mo–1V, two forms of annealing treatment are possible, mill annealing and duplex annealing. Mill annealing, in the case of this alloy, consists of heating at 790°C for 8 hours and then furnace cooling. Duplex annealing involves reheating the mill annealed material at 790°C for a quarter of an hour and then air cooling.

Table 8.2 Annealing treatment for titanium alloys

Alloy	Annealing		
	temp. (°C)	time (h)	cooling
Commercially pure Ti	650–760	0.1–2	air
Alpha	720–845	0.2–4	air
Near alpha	790–900	0.5–8	air
Alpha–beta	650–820	0.5–4	air/furnace
Beta	700–815	0.1–1	air/water

Solution and precipitation treatment

Table 8.3 shows typical solution treatment and aging processes for titanium alloys.

Table 8.3 Solution and precipitation treatment for titanium alloys

Alloy	Solution tr.			Precipitation	
	temp. (°C)	time (h)	cooling	temp. (°C)	time (h)
Alpha					
Commercially pure Ti			not used		
Ti–2.5Cu	805–815		air	475	8+
				400	8
Near alpha					
Ti–8Al–1Mo–1V	980–1010	1	oil/water	565–595	
Ti–6Al–2Sn–4Zr–2Mo	955–980	1	air	595	8
Ti–6Al–5Zr–0.5Mo–0.25 Si	1050		oil	550	24
Alpha–beta					
Ti–6Al–4V	950–970	1	water	480–595	4–8
Ti–4Al–4Mo–2Sn–0.5Si	900		air	500	24
Ti–6Al–6V–2Sn (Cu + Fe)	885–910	1	water	480–595	4–8
Ti–6Al–2Sn–4Zr–6Mo	845–890	1	air	580–605	4–8
Ti–5Al–2Sn–2Zr–4Mo–4Cr	845–870	1	air	580–605	4–8
Ti–6Al–2Sn–2Zr–2Mo–2 Cr–0.25Si	870–925	1	water	480–595	4–8
Beta or near beta					
Ti–13V–11Cr–3Al	775–800	0.25–1	air/water	425–480	4–100
Ti–11.5Mo–6Zr–4.5Sn	690–790	0.13–1	air/water	480–595	8–32
Ti–3Al–8V–6Cr–4Mo–4Zr	815–925	1	water	455–540	8–24

Stress relief

Table 8.4 shows typical stress relief temperatures and times for titanium alloys.

Table 8.4 Stress relief treatment for titanium alloys

Alloy	Stress relief		
	temp. (°C)	time (h)	cooling
Commercially pure Ti	480–600	0.25–1	air
Alpha	540–650	0.25–4	air
Near-alpha	480–700	0.25–4	air
Alpha-beta	480–700	0.25–4	air
Beta	675–760	0.1–0.5	air

8.4 Properties

Creep properties

The upper service temperature limit of near alpha, alpha-beta and beta titanium alloys is set by creep. The limit for commercially pure titanium and alpha alloys is set by the reduction in tensile strength. In general, the upper limit for commercially pure titanium is about 200°C, for other alpha alloys about 250 to 350°C, for near alpha about 500°C, for alpha-beta about 350 to 400°C. and for beta about 300°C. See the table with Mechanical properties for details of tensile strength at elevated temperatures.

Table 8.5 shows creep rupture data for commercially pure titanium at room and elevated temperatures.

Table 8.5 Creep rupture stresses for commercially pure titanium

Titanium	Temp. (°C)	Rupture stress MPa			
		100 h	1000 h	10 000 h	100 000 h
99.5Ti	20	286	269	255	241
	150	190	182	168	161
	300	139	134	128	122
99.2Ti	20	397	372	352	332
	150	253	250	244	239
	300	215	208	204	199
99.1Ti	20	455	423	392	357
	150	275	264	250	239
	300	216	215	205	199
99.0Ti	20	533	497	469	438
	150	321	296	281	269
	300	241	238	218	212

Density

The density of commercially pure titanium is 4.51×10^3 kg m^{-3} at 20°C. Alpha alloys have densities about 4.48×10^3 kg m^{-3}, near alpha 4.4 to 4.8×10^3 kg m^{-3}, alpha-beta 4.4 to 4.8×10^3 kg m^{-3}, and beta 4.8×10^3 kg m^{-3}.

Electrical resistivity

The electrical resistivity of commercially pure titanium at 20°C is 0.49 $\mu\Omega$ m, with alloys about 1.6 to 1.7 $\mu\Omega$ m.

Fatigue properties

The fatigue limit at 10^7 cycles for titanium is 0.50 times the tensile strength and for the alloys about 0.40 to 0.65 times the tensile strength.

Fracture toughness

Table 8.6 shows typical values of fracture toughness, K_{1c}, for a range of titanium alloys.

Table 8.6 Fracture toughness of titanium alloys

Alloy	Condition	K_{Ic} MPa m$^{-\frac{1}{2}}$
Alpha		
Commercially pure Ti	Annealed	>70
Ti–2.5Cu	Annealed	>70

Near alpha

Ti-6Al-5Zr-0.5Mo-0.25Si	Forging	60–70
Ti-6Al-2Sn-4Zr-2Mo	sol.tr, aged	50–60

Alpha–beta

Ti-6Al-4V	Annealed	50–60
Ti-4Al-4Mo-2Sn-0.5Si	Forging	45–55
	sol.tr, aged	40–50
Ti-4Al-4Mo-4Sn-0.5Si	sol.tr, aged	30–40

Beta

Ti-11.5Mo-6Zr-4.5Sn	sol.tr, aged	⟩50

Hardness

See Table 8.7 in Impact properties for typical hardness value. The absorption of oxygen by a titanium surface, when heated during manufacture, causes an increase in surface hardness.

Impact properties

Table 8.7 gives typical Charpy impact values, and hardness values, for a range of titanium and its alloys at 20°C.

Table 8.7 Impact and hardness data for titanium alloys

Alloy	Condition	Charpy (J)	Hardness HV	Hardness HB	Hardness HRC
Alpha–commercially pure					
99.5 Ti	Annealed			120	
99.2 Ti	Annealed	43		200	
99.1 Ti	Annealed	38		225	
99.0 Ti	Annealed	20		265	
99.2 Ti–0.2Pd	Annealed	43		200	
Alpha alloys					
Ti-5Al-2.5Sn	Annealed	26	220		36
Ti-5Al-2.5Sn (low oxygen)	Annealed	27	290		35
Near alpha alloys					
Ti-8Al-1Mo-1V	Duplex ann.	32			35
Ti-6Al-2Sn-4Zr-2Mo	Duplex ann.				32
Ti-6Al-1Mo-2Cb-1Ta	As rolled	31			30
Ti-6Al-5Zr-0.5Mo-0.2Si	Forging		355		
Alpha–beta alloys					
Ti-6Al-4V	Annealed	19	350		36
Ti-6Al-6V-2Sn	Annealed	18	365		38
Ti-7Al-4Mo	Sol + aged	18			38
Ti-4Al-4Mo-2Sn-0.5Si	Sol + aged		365		
Ti-4Al-4Mo-4Sn-0.5Si	Sol + aged		400		
Beta alloys					
Ti-11.5Mo-6Zr-4.5Sn	Sol + aged		400		
Ti-13V-11Cr-3Al	Sol + aged	11			40
Ti-8Mo-8V-2Fe-3Al	Sol + aged				40
Ti-3Al-8V-6Cr-4Mo-4Zr	Sol + aged	10			42

Machinability

Commercially pure titanium, alpha and near-alpha alloys have very good machinability. Some alpha-beta alloys have good machinability, e.g. Ti-6Al-4V, but others are only fair. Beta alloys have only fair machinability. Because of the relatively low thermal conductivity of titanium and its alloys, high local temperatures can be produced during machining and thus it is vital that cutting speed be carefully controlled.

Mechanical properties

The tensile modulus of titanium and its alloys is about 110 GPa, a notable exception being the alloy Ti–6Al–4V with a modulus of 125 GPa. The shear modulus is about 42–55 GPa, with most being 45 GPa. Table 8.8 shows the tensile properties of titanium and its alloys. See Impact properties for data on impact strength and hardness.

Table 8.8 Mechanical properties of titanium alloys

Alloy	Condition	Temp. (°C)	Tensile strength (MPa)	Yield stress (MPa)	Elong- ation (%)
Alpha–commercially pure					
99.5 Ti	Annealed	20	330	240	30
		315	150	95	32
99.2 Ti	Annealed	20	435	345	28
		315	195	115	35
99.1 Ti	Annealed	20	515	450	25
		315	235	140	34
99.0 Ti	Annealed	20	660	585	20
		315	310	170	25
99.2 Ti–0.2Pd	Annealed	20	435	345	28
		315	185	110	37
Alpha alloys					
Ti–5Al–2.5Sn	Annealed	20	860	805	16
		315	565	450	18
Ti–5Al–2.5Sn (low oxygen)	Annealed	20	805	745	16
		−255	1580	1420	15
Near-alpha alloys					
Ti–8Al–1Mo–1V	Duplex ann.	20	1000	950	15
		315	795	620	20
		540	620	515	25
Ti–6Al–2Sn–4Zr–2Mo	Duplex ann.	20	980	895	15
		315	770	585	16
		540	650	490	26
Ti–6Al–1Mo–2Cb–1Ta	As rolled	20	855	760	13
		315	585	460	20
		540	485	380	20
Ti–6Al–5Zr–0.5Mo–0.2Si	Forging	20	1040	875	10
		300	800	640	17
		500	690	500	19
Alpha-beta alloys					
Ti–6Al–4V	Annealed	20	990	925	14
		315	725	655	14
		540	530	425	35
	Sol + aged	20	1170	1100	10
		315	860	705	10
		540	655	485	22
Ti–6Al–6V–2Sn	Annealed	20	1070	1000	14
		315	930	805	18
	Sol + aged	20	1275	1170	10
		315	980	895	12
Ti–7Al–4Mo	Sol + aged	20	1105	1035	16
		315	975	745	18
Ti–4Al–4Mo–2Sn–0.5Si	Sol + aged	20	1190	960	17
		300	900	700	18

		500	790	600	21
Ti-4Al-4Mo-4Sn-0.5Si	Sol + aged	20	1330	1095	11
		300	1030	805	14
		500	900	690	17
Beta alloys					
Ti-11.5Mo-6Zr-4.5Sn	Sol + aged	20	1385	1315	11
		315	905	850	16
Ti-13V-11Cr-3Al	Sol + aged	20	1220	1170	8
		315	885	795	19
Ti-8Mo-8V-2Fe-3Al	Sol + aged	20	1310	1240	8
		315	1130	980	15
Ti-3Al-8V-6Cr-4Mo-4Zr	Sol + aged	20	1450	1380	7
		315	1035	895	20

Note: The yield strength values quoted above are for the 0.2% proof stress.

Thermal properties

The linear thermal expansivity, i.e. the linear coefficient of expansion, is about 8 to 9×10^{-6} °C^{-1} for titanium and its alloys. The specific heat capacity at 20°C is about 530 J kg^{-1} °C^{-1} for commercially pure titanium, and varies from about 400 to 650 J kg^{-1}°C^{-1} for titanium alloys. The thermal conductivity is about 16 W m^{-1} °C^{-1} for commercially pure titanium, and varies from about 5 to 12 W m^{-1} °C^{-1} for alloys.

Weldability

Commercially pure titanium and other alpha and near-alpha alloys generally have very good weldability. Some alpha-beta alloys are weldable, but beta are generally not. Among alpha-beta alloys, Ti-6Al-4V has good weldability. There is a need to guard against contamination, particularly from oxygen and nitrogen, at the high temperatures involved in the welding process. For this reason TIG welding is the most widely used process. Electron-beam, laser, plasma arc and friction welding processes can also be used. Resistance spot and seam welding is only used in situations where fatigue life is not important.

8.5 Uses

Forms

Table 8.9 shows the main forms available for titanium and its alloys.

Table 8.9 Forms of titanium alloys

Alloy	Forms						
	Bar	Forging	Sheet	Strip	Plate	Wire	Tube
Alpha–commercially pure							
99.5 Ti	*	*	*	*	*		
99.2 Ti	*	*	*	*	*		*
99.1 Ti	*	*	*	*	*	*	
99.0 Ti	*	*				*	
99.2 Ti–0.2Pd	*	*	*	*	*	*	

Alloy							
Alpha alloys							
Ti-5Al-2.5Sn	*	*		*	*	*	
Ti-5Al-2.5S (low oxygen)	*	*		*	*	*	
Near alpha alloys							
Ti-8Al-1Mo-1V	*	*					*
Ti-6Al-2Sn-4Zr-2Mo	*	*					
Ti-6Al-1Mo-2Cb-1Ta							*
Ti-6Al-5Zr-0.5Mo-0.2Si	*						
Alpha-beta alloys							
Ti-6Al-4V	*	*		*	*	*	*
Ti-6Al-6V-2Sn	*	*					
Ti-7Al-4Mo	*	*					
Ti-4Al-4Mo-2Sn-0.5Si	*	*				*	
Ti-4Al-4Mo-4Sn-0.5Si	*	*					
Beta alloys							
Ti-11.5Mo-6Zr-4.5Sn	*						*
Ti-13V-11Cr-3Al			*			*	
Ti-8Mo-8V-2Fe-3Al	*	*		*		*	*
Ti-3Al-8V-6Cr-4Mo-4Zr	*						*

Note: the term wire covers both wire and fastener stock.

Uses

Table 8.10 indicates typical uses of titanium and its alloys.

Table 8.10 Uses of titanium alloys

Alloy	Uses
Alpha-commercially pure	
99.5 Ti	Used where high ductility, good weldability and high corrosion resistance required, e.g. marine parts, airframes, chemical plant parts. Very good formability, machinability and forging properties.
99.2 Ti	As above.
99.1 Ti	As above.
99.0 Ti	As above., also used for gas compressors, high speed fans, aircraft engine parts. Highest strength commercially pure titanium.
99.2 Ti-0.2Pd	Has an improved resistance to corrosion in reducing media.
Alpha alloys	
Ti-5Al-2.5Sn	Combines the very good formability, machinability, weldability and forging properties of commercially pure titanium with improved strength at temperatures up to 350°C. Used for aircraft engine compressor blades and ducting, steam turbine blades.
Ti-5Al-2.5Sn (low oxygen)	A special grade for use at very low temperatures, down to-255°C.
Near-alpha alloys	
Ti-8Al-1Mo-1V	High strength, good creep resistance and toughness up to 450°C. Used for airframe and jet engine parts.
Ti-6Al-2Sn-4Zr-2Mo	A creep resistant alloy for use up to 475°C. Used for airframe skin

	components, parts and cases for jet engine compressors.
Ti–6Al–1Mo–2Cb–1Ta	High toughness, good resistance to sea water and hot-salt stress corrosion, good weldability with moderate strength.
Ti–6Al–5Zr–0.5Mo–0.2Si	Creep resistant alloy for use up to 550°C with good weldability.

Alpha–beta alloys

Ti–6Al–4V	Most widely used titanium alloy. Can be heat treated to different strength levels and has good weldability and machinability. Requires hot forming. Used for blades and discs for aircraft turbines and compressors, rocket motor cases, marine components, steam turbine blades, structural forgings and fasteners.
Ti–6Al–6V–2Sn	Used for structural aircraft parts and landing gear, rocket motor cases.
Ti–7Al–4Mo	Used for airframes and jet engine parts, missile forgings. Good properties up to 425°C.
Ti–4Al–4Mo–2Sn–0.5Si	A readily forged alloy with good creep resistance up to 400°C.
Ti–4Al–4Mo–4Sn–0.5Si	A forging alloy with very high strength at room temperature.

Beta alloys

Ti–11.5Mo–6Zr–4.5Sn	Can be cold formed. Used for high strength fasteners and aircraft sheet parts. Heat treatable.
Ti–13V–11Cr–3Al	Used for high strength fasteners, honeycomb panels, aerospace components. Heat treatable.
Ti–8Mo–8V–2Fe–3Al	Used for high strength fasteners, forged components and tough airframe sheets.
Ti–3Al–8V–6Cr–4Mo–4Zr	Used for high strength fasteners, aerospace components, torsion bars.

9 Polymeric materials

9.1 Materials

Types of polymer

Polymers can be grouped into three general categories:

1 Thermoplastics

These materials can be softened and resoftened indefinitely by the application of heat, provided the temperature is not so high as to cause decomposition. The term thermoplastic implies that the material becomes plastic when heated. Such polymers have linear or branched molecular chain structures with few links, if any, between chains. Linear, and some branched, polymer chains can become aligned so that a degree of crystallinity occurs.

2 Thermosets

These materials are rigid and not softened by the application of heat. Such polymers have molecular structures which are extensively cross-linked. Because of this, when heat causes bonds to break, the effect is not reversible on cooling.

3 Elastomers

Elastomers are polymers which, as a result of their molecular structure, allow considerable extensions which are reversible. Such materials are lightly cross-linked polymers. Between cross-links the molecular chains are fairly free to move. When stretched, the polymer chains tend to straighten and become aligned so that a degree of crystallinity occurs.

Elastomers

Elastomers can be grouped according to the form of their polymer chains.

1 Only carbon in the backbone of the polymer chain. This includes natural rubber, butadiene-styrene, butadiene acrylonitrile, butyl rubbers, polychloroprene and ethylene-propylene.

2 Polymer chains with oxygen in the backbone. Polypropylene oxide is an example.

3 Polymer chains with silicon in the backbone. Fluorosilicone is an example.

4 Polymer chains having sulphur in the backbone. Polysulphide is an example.

5 Thermoplastic elastomers. These are block copolymers with alternating hard and soft blocks. Examples are polyurethanes, ethylene vinyl acetate and styrene-butadiene-styrene. Such elastomers can be processed by thermoplastic moulding methods, such as injection and blow moulding. They, like thermoplastics, can be repeatedly softened by heating, unlike conventional elastomers.

See Additives, Crystallinity, Structure of polymers, Structure and properties, Codes, Composition, Chemical properties, Density, Glass transition temperature, Hardness, Mechanical properties, Permeability, Thermal properties, Uses.

Thermosets

The main thermosets are:

1 Phenolics
2 Amino resins (urea and melamine formaldehydes)
3 Epoxies
4 Unsaturated polyesters
5 Cross-linked polyurethanes

See Additives, Crystallinity, Structure of polymers, Structure and properties, Codes, Composition, Density, Electrical properties, Hardness, Mechanical properties, Thermal properties, Processing methods, Uses.

Thermoplastics

Thermoplastics can be considered as being two groups of material: ethnic and non-ethnic polymers.

1. The ethnic family of materials is based on ethylene. It can be subdivided into a number of other groups: polyolefins based on polyethylene and polypropylene; vinyls based on vinyl chloride, vinyl acetate and various other vinyl compounds. The backbone of all ethnic polymers is made up of just carbon atoms.

2. The non-ethnic family of materials has molecular backbones which include non-carbon atoms. Such polymers include polyamides, polyacetals, polycarbonates and cellulosics.

See Additives, Crystallinity, Structure of polymers, Structure and properties, Codes, Composition, Chemical properties, Creep properties, Density, Electrical properties, Fracture toughness, Glass transition temperature, Hardness, Impact properties, Mechanical properties, Optical properties, Permeability, Thermal properties, Processing methods, Uses.

9.2 Polymer structures

Additives

Plastics and rubbers almost invariably contain not only a polymeric material, but also additives. They may also be compounded from more than one polymer. The following are some of the main types of additive:

1. Fillers to modify the mechanical properties, to reduce for instance brittleness and increase the tensile modulus, e.g. wood flour, cork dust, chalk. They also have the effect of reducing the overall cost of the material.

2. Reinforcement, e.g. glass fibres or spheres, to improve the tensile modulus and strength.

3. Plasticizers to enable molecular chains to slide more easily past each other, hence making the material more flexible.

4. Stabilizers to enable the material to resist degradation better.

5. Flame retardants to improve the fire resistance properties.

6. Lubricants and heat stabilizers to assist the processing of the material.

7. Pigments and dyes to give colour to the material.

Crystallinity

Crystallinity is most likely to occur with polymers having simple linear chain molecules. Branched polymer chains are not easy to pack together in a regular manner, the branches get in the way. If the branches are regularly spaced along the chain, then some crystallinity is possible; irregularly spaced branches make crystallinity improbable. Heavily cross-linked polymers, i.e. thermosets, do not exhibit crystallinity but under stress some elastomers can. Table 9.1 shows the maximum possible crystallinity of some common polymers.

Table 9.1 Crystallinity of polymers

Polymer	Form of chain	Max. crystallinity
Polyethylene	Linear	95%
	Branched	60%
Polypropylene	Regularly spaced side groups on linear chain	60%
Polytetrafluoroethylene	Linear with bulky fluorine atoms in chain	75%
Polyoxymethylene	Linear with alternate oxygen and carbon atoms in chain	85%
Polyethylene terephthalate	Linear with groups in chain	65%
Polyamide	Linear with amide groups in chain	65%

Structure of polymers

Figure 9.1 shows the basic forms of a number of commonly used polymers. The figures are two dimensional representations of structures which, particularly in the case of thermosets and elastomers, are three dimensional.

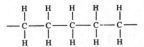

(a) Polyethylene, linear chain

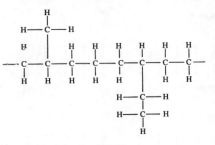

(b) Polyethylene, branched chain

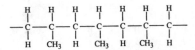

(c) Polypropylene, isotactic form (the main form)

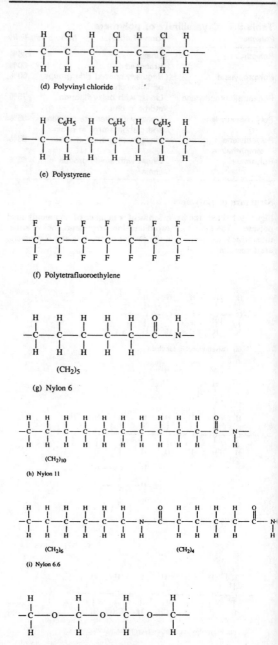

(d) Polyvinyl chloride

(e) Polystyrene

(f) Polytetrafluoroethylene

(g) Nylon 6

(h) Nylon 11

(i) Nylon 6.6

(j) Polyoxymethylene (acetal homopolymer)

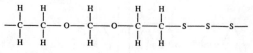

(k) Polysulphide

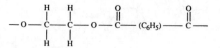

(l) Polyethylene terephthalate

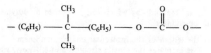

(m) Polycarbonate

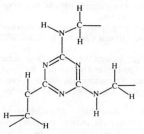

(n) Melamine formaldehyde

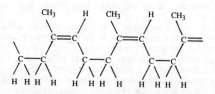

(o) Cis-polyisoprene, the natural rubber chain before vulcanization

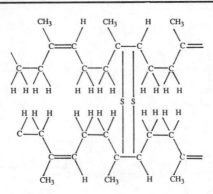

(p) Vulcanized rubber, sulphur linked polyisoprene chains

Figure 9.1 The basic forms of commonly used polymers

Structure and properties

The following are methods by which the properties of polymeric materials can be changed.

1 Increasing the length of the molecular chain for a linear polymer. This increases the tensile strength and stiffness, since longer chains more readily become tangled and so cannot so easily be moved.

2 Introducing large side groups into a linear chain. This increases the tensile strength and stiffness, since the side groups inhibit chain motion.

3 Producing branches on a linear chain. This increases the tensile strength and stiffness since the branches inhibit chain motion.

4 Introducing large groups into the chain. These reduce the ability of the chain to flex and so increase the rigidity.

5 Cross-linking chains. The greater the degree of cross-linking the more chain motion is inhibited and so the more rigid the material.

6 Introducing liquids between chains. The addition of liquids, termed plasticizers, which fill some of the space between polymer chains makes it easier for the chains to move and so increases flexibility.

7 Making some of the material crystalline. With linear chains a degree of crystallinity is possible. This degree can be controlled. The greater the degree of crystallinity the more dense the material, and the higher its tensile strength and stiffness.

8 Including fillers. The properties of polymeric materials can be affected by the introduction of fillers. Thus, for example, the tensile modulus and strength can be increased by incorporating glass fibres. Graphite as a filler can reduce friction.

9 Orientation. Stretching or applying shear stresses during processing can result in polymeric materials becoming aligned in a particular direction. The properties in that direction are then markedly different to those in the transverse direction.

10 Copolymerization. Combining two or more monomers in a single polymer chain will change the properties, the properties

being determined by the ratio of the components.
11 Blending. Mixing two or more polymers to form a material
will affect the properties, the result depending on the ratio of
the materials.

9.3 Codes and composition

Codes

Table 9.2 shows the common abbreviations and names used for
polymers. The full name used for a polymer aims, as much as is
possible, to describe the constitution of the primary polymer chain.
Thus for a homopolymer from a monomer X the polymer name is
polyX. If the monomer is XY then the polymer is polyXY. For
copolymers of monomers X and Y the polymer name is X–Y
copolymer or poly(X–co–Y). Not all polymers have names based
on the starting material, as above. Some have names based on the
repeat unit in the polymer chain. Thus if the name of the repeat unit
is X the polymer is called polyX. Others have names based on the
characteristic chemical unit common to the group of polymers, e.g.
epoxy.

Table 9.2 Abbreviations for polymers

Abbreviation	Common name	Polymer
Thermoplastics and thermosets		
ABS		acrylonitrile–butadiene–styrene
CA	acetate	cellulose acetate
CAB		cellulose acetate butyrate
CN	celluloid	cellulose nitrate
CPVC		chlorinated polyvinyl chloride
EP		epoxy, epoxide
MF	melamine	melamine formaldehyde
PA	nylon*	polyamide
PAN		polyacrylonitrile
PC		polycarbonate
PE	polythene	polyethylene
PETP	polyester	polyethylene terephthalate
PF	phenolic	phenol formaldehyde
PIB		polyisobutylene
PMMA	acryllic	polymethyl methacrylate
POM	acetal	polyoxymethylene
PP		polypropylene
PS	styrene	polystyrene
PTFE		polytetrafluorethylene
PUR		polyurethane
PVAC		polyvinyl acetate
PVAL		polyvinyl alcohol
PVC	vinyl	polyvinyl chloride
PVDC		polyvinylidene chloride
PVDF		polyvinylidene fluoride
PVF		polyvinyl fluoride
SAN		styrene–acrylonitrile
SB		styrene–butadiene
SI		silicone
UF	urea	urea formaldehyde
UP	polyester	unsaturated polyester
Elastomers		
ACM	acrylic	polyacrylate
ANM		acrylate–acrylonitrile copolymer
AU	polyurethane	polyester–urethane

BR		polybutadiene
BUTYL	butyl	isobutene–isoprene copolymer
CM		chlorinated polyethylene
CR	neoprene	polychloroprene
EPDM		ethylene–propylene–diene terpolymer
EPM		ethylene–propylene copolymer
EVA		ethylene vinyl acetate
EU	polyurethane	polyether–urethane
FKM	fluorocarbon	carbon chain fluoropolymer
FVMQ		fluorosilicone rubber
GPO		propylene oxide copolymer
IIR	butyl	isobutene–isoprene copolymer
NBR	Buna N, nitrile	acrylonitrile–butadiene copolymer
NR		natural rubber
SBR	GR–S, Buna S	styrene–butadiene copolymer
VMQ	silicone	polydimethylsiloxane, usually copolymer with vinyl groups
YSBR		thermoplastic styrene–butadiene copolymer

Note: * There are a number of common polyamides: nylon 6, nylon 6.6, nylon 6.10 and nylon 11. The numbers refer to the numbers of carbon atoms in each of the reacting substances used to produce the polymer (see Figure 9.1).

Composition

Most engineering polymeric materials include additives as well as the polymer, and some are compounded from more than one polymer. Suppliers of raw materials almost invariably supply a polymer range of different additives, the choice of combination depending on the purpose and hence properties required of the material. Thus, for example, one supplier of ABS has available nineteen different forms of the material. Some forms are just changes in colour as a result of different pigments being used as additives. Some versions have different percentages by weight of glass fibres and thus different mechanical properties. Some versions include fire retardants. Other forms are designed to have different impact properties as a result of the way the polymer chains are structured.

9.4 Properties

Chemical properties

Table 9.3 shows the chemical stability of thermoplastics at 20°C when exposed to different chemicals. Table 9.4 shows similar data for elastomers.

Table 9.3 Chemical stability of thermoplastics

Polymer	Water absorption	Acids		Alkalis		Organic solvents
		weak	strong	weak	strong	
ABS	M	R	AO	R	R	A
Acrylic	M	R	AO	R	R	A
Cellulose acetate	H	R	A	R	A	A
Cellulose acetate butyrate	H	R	A	R	A	A
Nylon	H	A	A	R	R	R
Polyacetal, copolymer	M	A	A	R	R	R

Polymer						
Polyacetal, homopolymer	M	R	A	R	A	R
Polycarbonate	L	R	A	A	A	A
Polyester	L	R	A	R	A	A
Polyethylene, high density	L	R	AO	R	R	R
Polyethylene, low density	L	R	AO	R	R	R
Polypropylene	L	R	AO	R	R	R
Polystyrene	L	R	AO	R	R	A
PTFE	L	R	R	R	R	R
PVC, unplasticized	M	R	R	R	R	A

Note: For water absorption L = low, less than 0.1% by weight in 24 hours immersion, M = medium, between 0.1 and 0.4%, H = high, more than 0.4% and often about 1%. R = resistant, A = attacked with AO being attack by oxidizing acids.

Table 9.4 Chemical stability of elastomers

Polymer	Acid	Alkali	Hydrocarbons (petrol, oil)	Chlorinated solvents	Oxidization	Ozone
Butadiene-acrylonitrile	G	G	E	P	F	P
Butadiene-styrene	G	G	P	P	G	P
Butyl	E	E	P	P	E	E
Chlorosulphonated polyethylene	G	G	G	P	E	E
Ethylene-propylene	E	E	P	P	E	E
Fluorocarbon	E	E	E	E	E	E
Natural rubber	G	G	P	P	G	P
Polychloroprene (neoprene)	G	G	G	P	E	E
Polysulphide	F	G	E	G	E	E
Polyurethane	F	P	E	P	E	E
Silicone	G	E	F	P	E	E

Note: E = excellent resistance, G = good, F = fair, P = poor.

Creep properties

Creep can be significant in polymeric materials at normal temperatures, the creep behaviour depending on stress and temperature as well as the type of material concerned. Generally, flexible polymeric materials show more creep than stiff ones.

The data obtained from creep tests is strain against time for a number of different stresses. From such data, a graph of stress against strain can be produced for different times, such a graph being known as an isochronous stress–strain graph. For a specific time, the quantity obtained from the graph by dividing the stress by the strain, a form of secant modulus, can be obtained. The result is known as the creep modulus. It is not the same as the tensile modulus, though it can be used to compare the stiffness of polymeric materials. The creep modulus varies both with time and strain. Table 9.5 shows some typical values of creep modulus for thermoplastics after one year under constant load at 20°C.

Table 9.5 Creep modulus for thermoplastics

Material	Creep modulus (GPa)
ABS	1.12
Acrylic	1.43
Ethylene–propylene copolymer	0.24
Nylon 6 (RH = 50%)	0.51
Nylon 66 (dry)	0.81
Nylon 66 (RH = 50%)	0.47
Nylon 66 + 33% glass fibres (RH = 50%)	3.5
Polyacetal	0.92
Polycarbonate	1.48
Polyester	1.30
Polyester, glass filled	2.8
Polyethylene, high density	0.10
Polyethylene, low density	0.29
Polypropylene	0.36
Polypropylene + 20% glass fibres	1.2
Polysulphone	2.03
PVC, unplasticized	1.51

Density

Table 9.6 shows the range of densities that occur with polymeric materials at 20°C.

Table 9.6 Densities of polymeric materials

Material	Density kg m^{-3}
Thermoplastics	
ABS	1020–1070
Acrylic	1180–1190
Cellulose acetate	1220–1340
Cellulose acetate butyrate	1150–1220
Polyacetal	1410–1420
Polyamide, Nylon 6 (dry)	1130–1140
Polyamide, Nylon 66 (dry)	1140–1150
Polycarbonate	1200
Polyester	1300–1350
Polyethylene, high density	935–970
Polyethylene, low density	913–970
Polyethylene terephthalate	1300
Polypropylene	900–910
Polystyrene	1050–1070
Polystyrene, toughened	1040–1060
Polysulphone	1240
Polytetrafluoroethylene	2140–2200
Polyurethane	1050–1250
Polyvinylchloride, unplasticized	1150–1400
Thermosets	
Epoxy, cast	1150
Epoxy, 60% glass fibre	1800
Melamine formaldehyde, cellulose	1500–1600
Phenol formaldehyde, unfilled	1250–1300
Phenol formaldehyde, wood flour	1320–1450
Phenol formaldehide, asbestos	1600–1850
Polyester, unfilled	1300
Polyester, 30% glass fibre	1500
Urea formaldehyde, cellulose fill.	1500–1600

Elastomers	
Butadiene-acrylonitrile	1000
Butadiene-styrene	940
Butyl	920
Chlorosulphonated polyethylene	1120–1280
Ethylene-propylene	860
Ethylene vinyl acetate	920–950
Fluorocarbon	1850
Natural rubber	930
Polychloroprene (neoprene)	1230
Polypropylene oxide	830
Polysulphide	1350
Polyurethane	1100–1250
Silicone	980
Styrene-butadiene-styrene	940–1030

Electrical properties

Table 9.7 shows the electrical resistivity (often referred to as the volume resistivity) and the relative permittivity (often referred to as the dielectric constant) of typical polymeric materials. It needs, however, to be recognised that the electrical properties can be markedly affected by the additives used with a polymer.

Table 9.7 Electrical properties of polymers

Polymer	Resistivity ($\mu\Omega$ m)	Relative permittivity
Thermoplastics		
ABS	10^{14}	2.4–2.5
Acrylic	10^{12}–10^{14}	2.2–3.2
Cellulose acetate	10^{8}–10^{12}	3.4–7.0
Cellulose acetate butyrate	10^{9}–10^{13}	3.4–6.4
Nylon 6 (moisture conditioned)	10^{10}–10^{13}	4.0–4.9
Nylon 66 (moisture conditioned)	10^{12}–10^{13}	3.9–4.5
Polyacetal	10^{13}	3.7
Polycarbonate	10^{14}	3.0
Polyethylene	10^{14}–10^{18}	2.3
Polypropylene	10^{13}–10^{15}	2.0
Polystyrene	10^{15}–10^{19}	2.4–2.7
Polysulphone	10^{14}–10^{17}	3.1
PVC, unplasticized	10^{14}	3.0–3.3
PVC, plasticized	10^{9}–10^{13}	4–8
Thermosets		
Melamine formaldehyde, cellulose	10^{10}	7.7–9.2
Phenolic, cellulose filler	10^{7}–10^{11}	4.4–9.0
Elastomers		
Butadiene-acrylonitrile	10^{8}	13
Butyl	10^{15}	2.1–2.4
Chlorosulphonated polyethylene	10^{12}	7–10
Ethylene-propylene	10^{14}	3.0–3.5
Natural rubber	10^{13}–10^{15}	2.3–3.0
Polychloroprene (neoprene)	10^{9}	9
Polyurethane	10^{10}	5–7
Silicone	10^{9}–10^{15}	3.0–3.5

Fracture toughness

Table 9.8 gives typical values of the plane strain fracture toughness in air of some thermoplastic materials at 20°C.

Table 9.8   Fracture toughness of thermoplastics

Material	K_{lc} MPa $m^{-\frac{1}{2}}$
Acrylic, cast sheet	1.6
Nylon 6.6	2.5–3.0
Polycarbonate	2.2
Polyethylene	1–6
Polypropylene	3.0–4.5
Polystyrene, general purpose	1.0
Polyvinyl chloride	2–4

Glass transition temperatures

Table 9.9 shows the glass transition temperatures of thermoplastics and elastomers. Also the normal condition of the material is given at 20°C.

Table 9.9   Glass transition temperatures of thermoplastics

Material	State	T_g °C
Thermoplastics		
ABS	amorphous	100
Acrylic	amorphous	100
Cellulose acetate	amorphous	120
Cellulose acetate butyrate	amorphous	120
Polyacetal, homopolymer	semi cryst.	−76
Polyamide, Nylon 6	semi cryst.	50
Polyamide, Nylon 66	semi cryst.	66
Polycarbonate	amorphous	150
Polyethylene, high density	semi cryst.	−120
Polyethylene, low density	semi cryst.	−90
Polyethylene terephthalate	semi cryst.	69
Polypropylene	semi cryst.	−10
Polystyrene	amorphous	100
Polysulphone	amorphous	190
Polytetrafluoroethylene	semi cryst.	−120
Polyvinylchloride	amorphous	87
Polyvinylidene chloride	semi cryst.	−17
Elastomers		
Butadiene-acrylonitrile	amorphous	−55
Butadiene-styrene	amorphous	−55
Butyl	amorphous	−79
Chlorosulphonated polyethylene	amorphous	−55
Ethylene–propylene	amorphous	−75
Ethylene vinyl acetate	amorphous	−160
Natural rubber	amorphous	−70
Polychloroprene (neoprene)	amorphous	−50
Polysulphide	amorphous	−50
Polyurethane	amorphous	−60
Silicone	amorphous	−50

Hardness

Table 9.10 shows typical hardness values at 20°C for polymeric materials. The main scales of hardness used with such materials are Rockwell and Shore. Hardness values are affected by stress, time under load and temperature.

Table 9.10 Hardness values for polymeric materials

Polymer	Hardness	
	Rockwell	Shore
Thermoplastics		
ABS	R90–115	
Acrylic	M90	
Cellulose acetate	R34–125	
Cellulose acetate butyrate	R31–116	
Polyacetal	M80–92, R115–120	
Polyamide, Nylon 6 (dry)	R120	
Polyamide, Nylon 66 (dry)	M80, R120	
Polycarbonate	M85, R120	
Polyester	M70–85	
Polyethylene, high density		D60–70
Polyethylene, low density		D40–51
Polypropylene	M70–75, R75–95	
Polystyrene	M70–80	
Polystyrene, toughened	M40–70	
Polysulphone	R120	
Polyvinylchloride, unplasticized	M115	D65–85
Polyvinylchloride, plasticized		A40–100
Thermosets		
Epoxy, glass filled	M100–112	
Melamine formaldehyde, cellulose	M115–125	
Phenol formaldehyde, cellulose	E64–95	
Elastomers		
Butadiene–acrylonitrile		A30–100
Butadiene–styrene		A40–100
Butyl		A30–100
Chlorosulphonated polyethylene		A50–100
Ethylene–propylene		A30–100
Fluorocarbon		A60–90
Natural rubber		A20–100
Polychloroprene (neoprene)		A20–100
Polyurethane		A20–100
Silicone		A30–80

Impact properties

Thermoplastics can be grouped into three groups according to their impact properties at 20°C.

1 Brittle, test specimens break even when unnotched. Acrylics, glass-filled nylon, polystyrene.
2 Notch brittle, test specimens do not break if unnotched but break when notched. ABS (some forms), acetals, acrylics (toughened), cellulosics, nylon (dry), polycarbonate (some forms), polyethylene (high density), polyethylene terephthalate, polypropylene, polysulphone, polyvinylchloride.
3 Tough, test specimens do not break even when sharp notched. ABS (some forms), nylon (wet), polycarbonate (some forms), polyethylene (low density), propylene–ethylene copolymer, PTFE.

Mechanical properties

Table 9.11 shows the mechanical properties of thermoplastics, thermosets and elastomers at 20°C, together with their approximate maximum continuous use temperature.

Table 9.11 Mechanical properties of polymers

Polymer	Tensile strength (MPa)	Tensile modulus (GPa)	Elongation (%)	Max. temp. (°C)
Thermoplastics				
ABS	17–58	1.4–3.1	10–140	70
Acrylic	50–70	2.7–3.5	5–8	100
Cellulose acetate	24–65	1.0–2.0	5–55	70
Cellulose acetate butyrate	18–48	0.5–1.4	40–90	70
Polyacetal, homopolymer	70	3.6	15–75	100
Polyamide, Nylon 6	75	1.1–3.1	60–320	110
Polyamide, Nylon 66	80	2.8–3.3	60–300	110
Polycarbonate	55–65	2.1–2.4	60–100	120
Polyethylene, high density	22–38	0.4–1.3	50–800	125
Polyethylene, low density	8–16	0.1–0.3	100–600	85
Polyethylene terephthalate	50–70	2.1–4.4	60–100	120
Polypropylene	30–40	1.1–1.6	50–600	150
Polystyrene	35–60	2.5–4.1	2–40	70
Polystyrene, toughened	17–24	1.8–3.1	8–50	70
Polysulphone	70	2.5	50–100	160
Polytetrafluoroethylene	14–35	0.4	200–600	260
PVC, unplasticised	52–58	2.4–4.1	2–40	70
PVC, low plasticiser	28–42		200–250	100
Thermosets				
Epoxy, cast	60–100	3.2		
Epoxy, 60% glass fibre	200–420	21–25		200
Melamine form., cellulose f.	55–85	7.0–10.5	0.5–1	95
Phenol formaldehyde	35–55	5.2–7.0	1–1.5	120
Phenol formaldehyde, wood flour filler	40–55	5.5–8.0	0.5–1	150
Phenol formaldehyde, asbestos filler	30–55	0.1–11.5	0.1–0.2	180
Polyester, unfilled	55	2.4		200
Olyester, 30% glass fibre	120	7.7	3	
Urea form., cellulose fill.	50–80	7.0–13.5	0.5–1	80
Elastomers				
Butadiene–acrylonitrile	28		700	100
Butadiene–styrene	24		600	80
Butyl	20		900	100
Chlorosulphonated polyethylene	21		500	130
Ethylene–propylene	20		300	100
Ethylene vinyl acetate	19		750	120
Fluorocarbon	18		300	230
Natural rubber	20		800	80
Polychloroprene (neoprene)	25		1000	100
Polypropylene oxide	14		300	170
Polysulphide	9		500	80
Polyurethane	40		650	80
Silicone	10		700	300
Styrene–butadiene–styrene	14		700	80

Optical properties

Table 9.12 gives the refractive index and the direct transmission factor for polymeric materials. This transmission factor is the percentage of light that would be transmitted if the material was 1 mm thick. Thinner material would give a higher transmission, thicker material would give less. Thus a transmission factor of 0% for 1 mm thickness does not mean that the material is completely opaque for thinner sheets.

Table 9.12 Optical properties of polymers

Polymer	Refractive index	Transmission factor (%)
Acetal copolymer	1.49	0
Acrylics	1.49	>99
Cellulose acetate	1.46–1.50	
Nylon 66	1.45	0
Polycarbonate	1.59	
Polyethylene, low density	1.52	45
Polyethylene terephthalate		>90
Polymethylepentene	1.47	99
Polypropylene	1.49	11–46
Polystyrene	1.59–1.60	
Polysulphone	1.63	
Polyvinylchloride	1.54	94
Polytetrafluoroethylene	1.32	0

Permeability

A common use of polymeric materials is as a barrier to gases and vapours. A measure of the rate at which gases or vapours can permeate through a material is given by the permeability, the higher the permeability, the greater will be the flow rate through a polymer film. See permeability in Chapter 1. Table 9.13 gives permeability values for a number of gases/vapours at 25°C.

Table 9.13 Permeability of polymers

Polymer	Permeability 10^{-18} kg m N^{-1} s^{-1}		
	to oxygen	water	carbon dioxide
Nylon 6	0.48	1350	6.0
Polyethylene, high density	6.4	40	30
Polyethylene, low density	17.6	300	171
Polyethylene terephthalate	0.22	600	0.9
Polypropylene	6.4	170	30
Polystyrene	9.33	300	120
PVC, unplasticized	0.37	400	2.9
Natural rubber	112.0	7700	1110
Butyl rubber	5.9		45

Resilience

To illustrate the term resilience, the more resilient a rubber ball, the higher it will bounce up after being dropped from a fixed height onto the floor. Natural rubber, butadiene–styrene, polychloroprene, and ethylene propylene have good resilience; butadiene–acrylonitrile, butyl, fluorosilicone and polysulphide fair resilience; polyurethane poor resilience.

Thermal properties

Table 9.14 shows the linear thermal expansivity, i.e. linear coefficient of expansion, specific heat capacity and thermal conductivity of polymeric materials. The thermal conductivity of a polymer can be markedly reduced if it is made cellular or foamed. Data for such materials is included at the end of the table.

Table 9.14 Thermal properties of polymeric materials

Polymer	Thermal expans. $(10^{-5}\ ^\circ C^{-1})$	Specific ht. cap. $(kJ\ kg^{-1}\ ^\circ C^{1})$	Thermal conduct. $(W\ m^{-1}\ ^\circ C^{-1})$
Thermoplastics			
ABS	8–10	1.5	0.13–0.20
Acrylic	6–7		0.13–0.15
Cellulose acetate	8–18	1.5	0.13–0.20
Cellulose acetate butyrate	11–17	1.5	0.13–0.20
Polyacetal, homopolymer	10	1.5	0.17
Polyamide, Nylon 6 (dry)	8–10	1.6	0.17–0.21
Polyamide, Nylon 66 (dry)	8–10	1.7	0.17–0.21
Polycarbonate	4–7	1.3	0.14–0.16
Polyethylene, high density	11–13	2.3	0.31–0.35
Polyethylene, low density	13–20	1.9	0.25
Polyethylene terephthalate		1.0	0.14
Polypropylene	10–12	1.9	0.16
Polystyrene	6–8	1.2	0.12–0.13
Polystyrene, toughened	7–8	1.4	0.12–0.13
Polysulphone	5	1.3	0.19
Polytetrafluoroethylene	10	1.1	0.27
PVC, unplasticized	5–19	1.1	0.12–0.14
PVC, low plasticizer	7–25	1.7	0.15
Thermosets			
Epoxy, cast	6	1.1	0.17
Epoxy, 60% glass fibre	1–5	0.8	
Melamine form., cellulose f.	4		0.27–0.42
Phenol formaldehyde	3–4		
Phenol formaldehyde, cellulose filler	3–4	1.5	0.16–0.32
Polyester, unfilled	6		
Elastomers			
Ethylene–propylene	6	1.1	0.17
Natural rubber		1.9	0.18
Polychloroprene (neoprene)	24	1.7	0.21
Cellular polymers			
Polystyrene, density 16 kg m⁻³			0.039
Polystyrene, density 32 kg m⁻³			0.032
Polyurethane, density 16 kg m⁻³			0.040
Polyurethane, density 32 kg m⁻³			0.023
Polyurethane, density 64 kg m⁻³			0.025

9.5 Uses

Processing methods

The forms available for finished polymeric material products depend on the processing methods that are possible. Thus casting can be used to produce sheets, rods, tubes and simple shapes. Extrusion is used for sheets, rods, tubes and long shapes. Injection moulding, compression moulding and transfer moulding are used for intricate shapes. Thermoforming is used to produce shapes from sheet material. Table 9.15 shows the processing methods possible with thermoplastics and Table 9.16 with thermosets.

Table 9.15 Processing methods for thermoplastics

Polymer	Extrusion	Injection mould	Extrusion blow mould	Rotational mould	Thermoform	Casting	Bend and join	As film
ABS	*	*		*	*		*	
Acrylic	*	*			*	*	*	
Cellulosics	*	*			*			*
Polyacetal	*	*	*					
Polyamide (nylon)	*	*		*		*	*	
Polycarbonate	*	*	*		*		*	
Polyester	*	*						
Polyethylene, high d.	*	*	*	*	*		*	*
Polyethylene, low d.	*	*	*	*			*	*
Polyethylene terephth.	*	*						*
Polypropylene	*	*	*		*		*	*
Polystyrene	*	*	*	*	*			*
Polysulphone	*	*			*			
PTFE	*							
PVC	*	*	*	*	*		*	*

Table 9.16 Processing methods with thermosets

Polymer	C or T moulding	Casting	Laminate	Foam	Film
Epoxy		*	*	*	
Melamine formaldehyde	*		*		
Phenol formaldehyde	*	*	*	*	
Polyester	*	*	*		*
Urea formaldehyde	*		*		

Note: C or T moulding is compression or transfer moulding.

Uses

Table 9.17 gives examples of the end products possible with polymeric materials.

Table 9.17 Uses of polymeric materials

Polymer	Uses
Thermoplastics	
ABS	Tough, stiff and abrasion resistant. Used as casings for telephones, vacuum cleaners, hair driers, TV sets, radios, typewriters, luggage, boat shells, food containers.
Acrylic	Transparent, stiff, strong and weather resistant. Used for light fittings, canopies, lenses for car lights, signs and nameplates. Opaque sheet is used for domestic baths, shower cabinets, basins, lavatory cisterns.

Cellulose acetate	Hard, stiff, tough but poor dimensional stability. Used for spectacle frames, tool handles, toys, buttons.
Cellulose acetate butyrate	More resistant to solvent attack, lower moisture absorption, higher impact strength than cellulose acetate. Used for cable insulation, natural gas pipes, street signs, street light globes, tool handles, lenses for instrument panel lights, blister packaging, containers.
Polyacetal, homopolymer	Stiff, strong and maintains properties at relatively high temperatures. Used as pipe fittings, parts for water pumps and washing machines, car instrument housings, bearings, gears, hinges, window catches, car seat belt buckles.
Polyamide, Nylon	Nylon 6 and 66 are the most widely used, 66 having a higher melting point and being stronger and stiffer than 6 but more absorbent of water. Used for gears, bearings, bushes, housings for power tools, electric plugs and sockets, and as fibres in clothing.
Polycarbonate	Tough, stiff and strong. Used where impact resistance and relatively high temperatures experienced. Used for street lamp covers, infant feeding bottles, machine housings, safety helmets, cups and saucers.
Polyethylene, high density	Good chemical resistance, low moisture absorption and high electrical resistance. Used for piping, toys, household ware.
Polyethylene, low density	Good chemical resistance, low moisture absorption and high electrical resistance. Used for bags, squeeze bottles, ball-point pen tubing, wire and cable insulation.
Polyethylene terephthalate	In fibre form used for clothes. Used for electrical plugs and sockets, wire insulation, recording tapes, insulating tape, gaskets, and widely as a container for fizzy drinks.
Polypropylene	Used for crates, containers, fans, car fascia panels, tops of washing machines, radio and TV cabinets, toys, chair shells.
Polystyrene	With no additives is brittle and transparent; blending with rubber gives a toughened form. This form is used as vending machine cups, casings for cameras, projectors, radios, TV sets and vacuum cleaners. Is excellent electrical insulator and used in electrical equipment. Foamed, or expanded,

	polystyrene is used for insulation and packaging.
Polysulphone	Strong, stiff and excellent creep resistance. Burns with difficulty and does not present a smoke hazard. Used in aircraft as parts in passenger service units, circuit boards, coil bobbins, circuit- breaker items, cooker control knobs.
Poly tetrafluoroethylene	Tough and flexible, can be used over very wide temperature range. Used for piping carrying corrosive chemicals, gaskets, diaphragms, valves, O-rings, bellows, couplings, dry and self-lubricating bearings, and because other materials will not bond with it, coatings for non-stick pans, coverings for rollers, linings for hoppers and chutes.
PVC, unplasticized	Rigid, used for piping for waste and soil drainage systems, rain water pipes, lighting fittings, curtain rails.
PVC, plasticized	Flexible, used for plastic rain coats, bottles, shoe soles, garden hose pipes, gaskets, inflatable toys.
Thermosets	
Epoxy	Used as an adhesive, a laminating resin, and as a coating for structural steel, masonry and marine items.
Epoxy, cast	Encapsulation of electronic components, for fabrication of short run moulds, patterns.
Epoxy, with glass fibres	Gives hard, strong composites. Used for boat hulls and table tops.
Melamine formaldehyde	Supplied as a moulding powder, this includes the resin, fillers and pigments. Used for cups and saucers, knobs, handles, light fittings, toys. Composites with open-weave fabrics used for building panels and electrical equipment.
Phenol formaldehyde	Also known as Bakelite. Supplied as a moulding powder, this includes resin, fillers and pigments. Fillers account for 50 to 80% by weight of the moulding powder. Wood flour increases impact strength, asbestos fibres increase heat properties, mica increases electrical resistance. Used for electrical plugs and sockets, switches, door knobs and handles, camera bodies, ashtrays. Composites with paper or an open-weave fabric are used for gears, bearings, and electrical insulation parts.
Polyester	Generally used as a composite with glass fibres. Used for boat hulls, building panels, stackable chairs.
Urea formaldehyde	As melamine formaldehyde.

Elastomers

Butadiene–acrylonitrile	Excellent resistance to organic liquids. Used for hoses, gaskets, seals, tank linings, rollers.
Butadiene–styrene	Cheaper than natural rubber. Used in the manufacture of tyres, hosepipes, conveyor belts, cable insulation.
Butyl	Extremely impermeable to gases. Used for inner linings of tubeless tyres, steam hoses, diaphragms.
Chlorosulphonated polyethylene	Trade name Hypalon. Excellent resistance to ozone with good chemical resistance abrasion, fatigue and impact properties. Used for flexible hose for oil and chemicals, tank linings, cable insulation, V-belts, O-rings, seals, gaskets, shoe soles.
Ethylene–propylene	Very high resistance to oxygen, ozone and heat. Good electrical and aging properties. Used for electric insulation, footwear, hose, belts.
Ethylene vinyl acetate	Flexible, good impact strength, high clarity. Used for cable insulation, flexible tubing, shoe soles, gaskets.
Fluorocarbon	Used for O-rings, seals, gaskets, diaphragms.
Natural rubber	Inferior to synthetics in oil and solvent resistance and oxidation resistance. Attacked by ozone. Higher tear resistance than synthetic rubbers. Used for tyres, gaskets, hose.
Polychloroprene (neoprene)	Good resistance to oils and good weathering characteristics. Used for oil and petrol hoses, gaskets, seals, diaphragms, chemical tank linings.
Polypropylene oxide	Excellent impact and tear strength with good mechanical properties and resilience. Used for electrical insulation.
Polysulphide	Often has an offensive odour. Excellent resistance to oils and solvents, low permeability to gases. Can be attacked by micro-organisms. Used for printing rolls, cable coverings, coated fabrics, sealants in building work.
Polyurethane	Widely used for both flexible and rigid foams. Used as cushioning, packaging, structural and insulation panels.
Silicone	Can be used over a wide temperature range.-90 to 250°C and higher for some grades. Chemically inert, good electrical properties, but expensive. Used for electric insulation seals, shock mounts, vibration dampers, adhesives.

Styrene–butadiene–styrene Known as a thermoplastic rubber. Properties controlled by ratio of styrene and butadiene. Properties comparable with natural rubber. Used for footwear, carpet backing, and in adhesives.

10 Ceramics

10.1 *Materials*

Ceramics

The term ceramics covers a wide range of materials, e.g. brick, stone, glasses and refractory materials. Ceramics are formed from combinations of one or more metals with a non-metallic element, such as oxygen, nitrogen or carbon. Ceramics are usually hard and brittle, good electrical and thermal insulators, and have good resistance to chemical attack. They tend to have a low thermal shock resistance, because of their low thermal conductivity and thermal expansivity.

Ceramics are usually crystalline, though amorphous states are possible. If, for instance, silica in the molten state is cooled very slowly it crystallizes at the freezing point. However, if it is cooled more rapidly it is unable to get its atoms into the orderly state required of a crystal and the resulting solid is a disorderly arrangement called a glass.

Engineering ceramics

Common engineering ceramics are alumina (an oxide of aluminium), boron carbide, silicon nitride, silicon carbide, tantalum carbide, titanium carbide, tungsten carbide, and zirconium carbide. A major use of such materials is as cemented tips for tools. They are bonded with a metal binder such as nickel, cobalt, chromium, or molybdenum to form a composite material. The most common forms are tungsten carbide bonded with cobalt and a more complex form involving a number of carbides with cobalt.

See Electrical properties, Mechanical properties of alumina ceramics, Mechanical properties of bonded ceramics, Thermal properties of bonded ceramics, Uses of alumina ceramics, Uses of bonded ceramics.

Glasses

The basic ingredient of most glasses is sand, i.e. the ceramic silica. Ordinary window glass is made from a mixture of sand, limestone (calcium carbonate) and soda ash (sodium carbonate). Heat resistant glasses, such as Pyrex, are made by replacing the soda ash by boric oxide. The tensile strength is markedly affected by microscopic defects and surface scratches. They have low ductility, being brittle, and have low thermal expansivity and low thermal conductivity, hence poor resistance to thermal shock. They are good electrical insulators and are resistant to many acids, solvents and other chemicals.

See Electrical properties, Mechanical properties of glasses, Thermal properties of glasses, Uses of glasses.

Refractories

These are special materials used in construction which are capable of withstanding high temperatures. One of the most widely used refractories consists of silica and alumina. Figure 10.1 shows the equilibrium diagram. The ability of a material from these materials to withstand high temperatures (the term refractoriness being used to describe this) increases with an increase in alumina above the eutectic point.

See Mechanical properties of alumina ceramics, Uses of alumina ceramics.

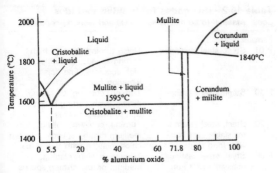

Figure 10.1 Equilibrium diagram for silica-alumina

10.2 Codes

Codes for bonded carbides

The American coding system for bonded carbides used for tools consists of the letter C followed by a number to indicate the machining characteristics of the carbide. A code, such as C-5, covers a number of carbides but indicates that each is capable of rough machining of carbon and alloy steels. Table 10.1 shows the main parts of the system. The ISO have a similar system, i.e. indicating what a carbide can do rather than its composition. They use a letter P, M or K, followed by two digits. Table 10.2 shows the system. The ISO group K roughly equates to the American codes C–1 to C–4, P to C–5 to C–8, with M giving an intermediate group.

Table 10.1 American codes for bonded carbides

Code	Machining applications
Cast iron, non-ferrous and non-metallic materials	
C–1	Roughing
C–2	General purpose
C–3	Finishing
C–4	Precision finishing
Carbon and alloy steels	
C–5	Roughing
C–6	General purpose
C–7	Finishing
C–8	Precision finishing
Wear-surface applications	
C–9	No shock
C–10	Light shock
C–11	Heavy shock
Impact applications	
C–12	Light impact
C–13	Medium impact
C–14	Heavy impact
Miscellaneous applications	
C–15 to C–19	

Table 10.2 ISO codes for bonded carbides

Code	Material to be machined	Use and working conditions
Ferrous metals		
P 01	Steel, steel castings	Finish turning, boring, high cutting speeds, thin chips, high accuracy, fine finish, vibration free.
P 10	Steel, steel castings	Turning, copying, threading, milling, high cutting speeds, thin to medium chips.
P 20	Steel, steel castings, malleable cast iron with long chips.	Turning, copying, milling, medium cutting speeds, medium chips, planing with thin chips.
P 30	Steel, steel castings, malleable cast iron with long chips.	Turning, milling, planing, medium or low cutting speeds, medium or large chips, use in unfavourable conditions*.
P 40	Steel, steel castings with sand inclusions and voids.	Turning, planing, slotting, low cutting speeds, thick chips, large cutting angles in unfavourable conditions* and on automatic machines.
P 50	Steel, medium or low tensile strength steel castings with sand inclusions and voids.	Operations requiring very tough carbide, turning, planing, slotting, low cutting speeds, thick chips, large cutting angles in unfavourable conditions* and on automatic machines.
Ferrous and non-ferrous metals		
M 10	Steel, steel castings, manganese steel, grey cast iron, alloy cast iron.	Turning, medium or high cutting speeds, thin to medium chips.
M 20	Steel, steel castings, austenitic or manganese steel, grey cast iron.	Turning, milling, medium cutting speeds, medium chips.
M 30	Steel, steel castings, austenitic steel, grey cast iron, high temperature alloys.	Turning, milling, medium cutting speeds, medium or thick chips.
M 40	Mild free cutting steel, low tensile steel, non-ferrous metals and light alloys.	Turning, cut-off, particularly on automatic machines.
Ferrous and non-ferrous metals, and non-metals		
K 01	Very hard grey cast iron, chilled castings, high silicon aluminium alloys, hardened steel, highly abrasive plastics, hard cardboard, ceramics.	Turning, finish turning, boring, milling, scraping.
K 10	Grey cast iron over 220 BH, malleable cast iron with short chips, hardened steel, silicon aluminium alloys, copper alloys, plastics, glass, hard rubber, hard	Turning, milling, drilling, boring, broaching, scraping.

	cardboard, porcelain, stone.	
K 20	Grey cast iron up to 220 BH, non-ferrous metals, copper, brass aluminium.	Turning, milling, planing, boring, broaching, demanding
K 30	Low hardness grey cast iron, low tensile steel, compressed wood.	Turning, milling, planing, slotting, use in unfavourable conditions* and with large cutting angles.
K 40	Soft wood, hard wood, non-ferrous metals.	Turning, milling, planing, slotting, use in unfavourable conditions*, and with large cutting angles.

Note: * Unfavourable conditions include shapes that are awkward to machine, material having a casting or forging skin, material having a variable hardness, and machining that involves variable depth of cut, interrupted cut or moderate to severe vibrations. In each of the above categories, the lower the number the higher the cutting speed and the lighter the feed, the higher the number the slower the cutting speed but the heavier the feed.

10.3 Properties

Density

See Mechanical properties for values.

Electrical properties

Most ceramics have resistivities greater than 10^{13} Ω m with glasses about 10^4 to 10^{10} Ω m. The permittivity, i.e. dielectric constant, of alumina ceramics is about 8 to 10 with glasses about 4 to 7.

Mechanical properties of alumina ceramics

Table 10.3 shows the mechanical properties of alumina–silica ceramics.

Table 10.3   Mechanical properties of alumina ceramics

Alumina (%)	Density 10^3 kg m^{-3}	Hardness Rockwell (A)	Compressive strength (MPa)	Tensile strength (MPa)	Elastic modulus (GPa)
85	3.39	73	1930	155	221
90	3.60	79	2480	220	276
96	3.72	78	2070	190	303
99.9	3.96	90	3790	310	386

Mechanical properties of bonded carbides

Table 10.4 shows the mechanical properties of commonly used bonded carbides. The properties depend on the grain size and so the data quoted relate to the spread of grain size commonly encountered. The modulus of elasticity is about 640 GPa with high tungsten carbide content (97%), dropping to about 480 GPa for 75% content.

Table 10.4 Mechanical properties of bonded carbides

Composition Main constituents (%)	Hardness Rockwell (A)	Density (10^3 kg m^{-3})	Compressive strength (MPa)	Impact strength (J)
97WC–3Co	92–93	15.0	5860	1.1
94WC–6Co	90–93	15.0	5170–5930	1.0–1.4
90WC–10Co	87–91	14.5	4000–5170	1.7–2.0
84WC–16Co	86–89	13.9	3860–4070	2.8–3.1
75WC–25Co	83–85	13.0	3100	3.1
71WC–12.5TiC–12TaC–4.5Co	92–93	12.0	5790	0.8
72WC–8TiC–11.5TaC–8.5Co	90–92	12.6	5170	0.9
64TiC–28WC–2TaC–2Cr$_2$C$_3$–5.0Co	94–95	6.6	4340	
57WC–27TaC–16Co	84–86	13.7	3720	2.0

Note: WC = tungsten carbide, TiC = titanium carbide, TaC = tantalum carbide, Cr$_{+2}$C$_3$ = chromium carbide, Co = cobalt.

Mechanical properties of glasses

The tensile strength of glasses is very much affected by microscopic defects and surface scratches and for design purposes a value of about 50 MPa is generally used. Glasses have a tensile modulus of about 70 GPa.

Thermal properties of bonded carbides

Table 10.5 gives the linear thermal expansivity and thermal conductivity of bonded carbides. The expansivities are quoted for temperatures of 200°C and 1000°C. At a particular temperature, a low expansivity and low thermal conductivity means that the material is susceptible to thermal shock if there are sudden changes in temperature.

Table 10.5 Thermal properties of bonded carbides

Composition Main constituents (%)	Thermal expansivity (10^{-6}°C^{-1})		Thermal conductivity (W m^{-1} °C^{-1})
	at 200°C	1000°C	
97WC–3Co	4.0		121
94WC–6Co	4.3	5.4–5.9	100–121
90WC–10Co	5.2		112
84WC–16Co	5.8	7.0	88
75WC–25Co	6.3		71
71WC–12.5TiC–12TaC–4.5Co	5.2	6.5	35
72WC–8TiC–11.5TaC–8.5Co	5.8	6.8	50
57WC–27TaC–16Co	5.9	7.7	

Note: WC = tungsten carbide, TiC = titanium carbide,

Table 10.6 Thermal properties of glasses

Glass	Thermal expansivity (10^{-7} °C^{-1})	Maximum service (temp. °C)
Alumina-silicate	42	910
Boro-silicate (Pyrex)	33	760
Fused silica (99.9%)	6	1470
(96%)	8	1500
Lead alkali (54% silica)	90	650
Soda-lime silica	92	730

10.4 Uses

Uses of alumina ceramics

A widely used refractory consists of alumina and silica. Refractoriness increases with an increase in alumina content. With about 20 to 40% alumina, the product finds use as fireclay refractory bricks. For more severe conditions, the amount of alumina is increased, with more than 71.8% alumina the ceramic can be used up to 1800°C. With 90% a tough, fine grained ceramic is produced for use in demanding mechanical conditions. With 96% a ceramic is produced which is excellent for special electronic applications. With 99.9% a hard, strong, ceramic is produced which is used in severe mechanical applications and hostile environments.

Uses of bonded carbides

Table 10.7 shows the types of tool applications for bonded carbides. Also see Codes for bonded carbides for details of uses in relation to code specification.

Table 10.7 Uses of bonded carbides

Composition (%)	Grain size	Tool applications
97WC–3Co	medium	Excellent abrasion resistance, low shock resistance, maintains sharp cutting edge. Used for machining cast iron, non-ferrous metals and non-metallic materials.
94WC–6Co	fine	Used for machining non-ferrous and high-temperature alloys.
	medium	Used for general purpose machining of metals other than steel, also small and medium compacting dies.
	coarse	Used for machining cast iron, non-ferrous metals and non-metallic materials, also small wire-drawing dies and compacting dies.
90WC–10Co	fine	Used for machining steel and milling high-temperature metals, form tools, face mills, end mills, cut-off tools, screw machine tools.
84WC–16Co	fine	Used for mining roller bits and percussive drilling bits.
	coarse	Used for medium and large dies

		where toughness is required, blanking dies and large mandrels.
75WC–25Co	medium	Used for heading dies, cold extrusion dies, punches and dies for blanking heavy stock.
71WC–12.5TiC–12TaC–4.5Co	medium	Used for finishing and light roughing work on plain carbon, alloy steels and alloy cast irons.
72WC–8TiC–11.5TaC–8.5Co	medium	Tough, wear resistant, withstands high temperatures. Used for heavy duty machining, milling plain carbon, alloy steels and alloy cast irons.
64TiC–28WC–2TaC–2Cr$_2$C$_3$–5.0Co	medium	Used for high speed finishing of steels and cast irons.
57WC–27TaC–16Co	coarse	Used for cutting hot flash from welded tubing, and dies for hot extrusion of aluminium.

Note: WC = tungsten carbide, TiC = titanium carbide, TaC = tantalum carbide, Cr$_2$C$_3$ = chromium carbide, Co = cobalt.

Uses of glasses

Table 10.8 shows the basic characteristics and uses of commonly used glasses.

Table 10.8   Uses of glasses

Glass	Uses
Alumina-silicate	Thermal shock resistant. Used for thermometers.
Boro-silicate (Pyrex)	Thermal shock resistant and easy to form. Used for glass cooking utensils.
Fused silica	Thermal shock resistant. Used for laboratory equipment.
Lead alkali (54% silica)	Has a high refractive index. Used for cut glass items.
Soda–lime silica	Easy to form. Used as plate glass, windows, bottles.

11 Composites

11.1 Materials

Types of composites

Composites can be classified into three main categories:

1 Fibre reinforced

Examples are vehicle tyres (rubber reinforced with woven cords), reinforced concrete, glass fibre reinforced plastics, carbon fibres in epoxy resins or aluminium, wood (a natural composite with tubes of cellulose in a matrix of lignin).

2 Particle reinforced

Examples are polymeric materials incorporating fillers, such as glass spheres or finely divided powders, toughened polymers in which fine rubber particles are included, cermets with ceramic particles in a metal matrix.

3 Dispersion strengthened

Examples are aluminium alloys following solution treatment and precipitation hardening, maraging steels, sintered metals.

In addition there is the entire range of laminated materials. Examples include plywood, clad metals, metal honeycomb structures, corrugated cardboard.

Fibre reinforced materials

For fibre reinforced composites, the main functions of the fibres are to carry most of the load applied to the composite and provide stiffness. For this reason, fibre materials have high tensile strength and a high tensile modulus. The properties required of the matrix material are that it adheres to the fibre surfaces so that forces applied to the composite are transmitted to the fibres since they are primarily responsible for the strength of the composite, that it protects the fibre surfaces from damage and that it keeps the fibres apart to hinder crack propagation. The fibres used may be continuous, in lengths running the full length of the composite, or discontinuous in short lengths. They may be aligned so that they are all in the same direction, so giving directionality to the properties, or randomly orientated.

For continuous fibres:

composite strength $= \sigma_f f_f + \sigma_m f_m$,
composite modulus $= E_f f_f + E_m f_m$,

where σ_f = stress on fibres, σ_m = stress on matrix, f_f = fraction of composite cross-section fibre, f_m = fraction matrix, E_f = tensile modulus of fibres, E_m = tensile modulus of matrix.

For discontinuous fibres the composite strength is given by the same equation but with the σ_f term replaced by an average stress term.

Average stress $= \sigma_{fu}(1 - L_c/2L)$,

where σ_{fu} = the maximum strength of the fibres, i.e. tensile strength, L = length of fibres and L_c is the critical length of the fibres. Provided the fibre length is equal to or greater than the critical length, the stress reaches the maximum value possible in the fibre. The critical length is given by

critical length $= \sigma_{fu} D/2\tau_m$,

where τ_m = the shear strength of the matrix, D = diameter of the fibres.

See Properties of fibre reinforced materials and Properties of wood.

Particle reinforced materials

Particle reinforced materials have particles of 1 μm or more in diameter dispersed throughout a matrix, the particles often accounting for a quarter to a half, or even more, of the total volume of the composite. Many polymeric materials incorporate fillers, the effect of the filler particles on the properties of the matrix material being generally to increase the tensile modulus, the tensile strength and impact resistance, reduce creep and thermal expansivity, and lower the overall cost of the material, since the polymer costs more than the filler.

The addition of spherical carbon black particles to rubbers is used to, amongst other things, improve the tensile modulus. The tensile modulus of the composite rubber is related to the modulus of the unfilled rubber E by

composite modulus $E(1 + 2.5f + 14.1f^2)$

where f is the volume fraction of the composite that is carbon.

The toughness of some polymers can be increased by incorporating fine rubber particles in the polymer matrix. For example, polystyrene is toughened by polybutadiene to give a product referred to as high-impact polystyrene (HIPS). Styrene–acrylonitrile is toughened with polybutadiene or styrene–butadiene copolymer to give acrylonitrile–butadiene–styrene terpolymer (ABS). The rubber has a lower tensile modulus than the matrix material and the net result is a lowering of the tensile modulus and tensile strength but much greater elongations before breaking, hence the improvement in toughness.

Cermets are composites involving ceramic particles in a matrix of metal. The ceramics have a high strength, high tensile modulus, high hardness, but are brittle. By comparison, the metals are weaker and less stiff, but ductile. The resulting composite is strong, hard and relatively tough. See Chapter 10 and bonded ceramics for more details of this type of material.

Dispersion strengthened metals

The strength of a metal can be increased by small particles dispersed throughout it. One way of doing this is solution treatment followed by precipitation hardening. Such treatments are used with, for example, maraging steels (see Chapter 3) and aluminium alloys (see Chapter 4).

Another way of introducing a dispersion of small particles throughout a metal, involves sintering. This process involves compacting a powdered metal powder in a die and then heating it to a temperature high enough to knit together the particles in the powder. If this is done with aluminium, the result is a fine dispersion of aluminium oxide, about 10%, throughout an aluminium matrix.

11.2 Properties

Properties of fibre reinforced materials

Table 11.1 shows the properties of commonly used fibres and whiskers. Whiskers differ from fibres in that they are grown as single crystals, rather than being polycrystalline as are fibres. Table 11.2 shows the properties of some fibre-reinforced materials. Table 11.3 shows some critical length data for discontinuous fibres in a matrix.

Table 11.1 Properties of fibres

Fibre	Density (10^3 kg m^{-3})	Tensile modulus (GPa)	Tensile strength (MPa)
Glasses			
E-glass	2.5	73	2.5
S-glass	2.5	86	4.6
Silica	2.2	74	5.9
Polycrystalline materials			
Alumina	3.2	173	2.1
Boron	2.6	414	2.8
Carbon	1.8	544	2.6
Silicon carbide	4.1	510	2.1
Whiskers			
Alumina	3.9	1550	20.8
Boron carbide	2.5	450	6.9
Graphite	2.2	700	20.7
Silicon carbide	3.2	700	21
Silicon nitride	3.2	380	7.0
Metals			
Molybdenum	10.2	335	2.2
Steel	7.7	200	4.2
Tungsten	19.3	345	2.9

Table 11.2 Properties of fibre reinforced materials

Composite	Density (10^3 kg m^{-3})	Tensile modulus (GPa)	Tensile strength (MPa)
Polymer matrix			
Epoxy + 14% alumina whiskers	1.6	42	800
Epoxy + 35% silicon nitride whiskers	1.9	105	280
Epoxy + 58% carbon	1.7	165	1520
Epoxy + 72% E glass	2.2	56	1640
Epoxy + 72% S glass	2.1	66	1900
Nylon 66 + 30% glass	1.4		140
Polyacetal + 20% glass	1.6		75
Polycarbonate + 20% glass	1.4		110
Polyethylene + 20% glass	1.1		40
Polypropylene + 20% glass	1.1		40
Polyester + 65% E glass	1.8	20	340
Metal matrix			
Aluminium + 50% boron	2.7	207	1140
Aluminium + 47% silica			910
Copper + 50% tungsten	14.1	262	1210
Copper + 77% tungsten			1800
Nickel + 8% boron			2700

Nickel + 40% tungsten	1100

Table 11.3 Critical length data for fibres in a matrix

Fibre	Fibre diameter (μm)	Matrix	Critical length (mm)
Alumina whiskers	2	Epoxy	0.5
Boron	100	Aluminium	1.8
Boron	100	Epoxy	3.5
Carbon	7	Epoxy	0.2
Glass	13	Epoxy	0.4
Glass	13	Polyester	0.5
Tungsten	2000	Copper	38

Properties of woods

Table 11.4 shows the approximate mechanical properties of common woods.

Table 11.4 Properties of woods

Wood	Density (10^3 kg m^{-3})	Elastic modulus —	Elastic modulus ∥	Tensile strength (MPA)	Compressive strength =	Compressive strength ∥
Ash	0.58	16	0.9	100	51	11
Beech	0.70	14	1.1	100		
Cedar, red	0.47	6		60	42	8
Elm	0.50	9		80	38	6
Mahogany	0.50	12	0.6	100		
Oak	0.65	11		100	49	9
Pine (white)	0.36	9		60	34	4
Spruce	0.40	11	0.6	70	40	5
Teak	0.60	13		100		

Note: — = parallel to grain, ∥ = perpendicular to grain.

12 Electrical properties

12.1 *Electrical conduction*

Energy bands

The electrons in isolated atoms occupy discrete energy levels. However, if the atoms are packed together to form a solid, an atom cannot now be considered in isolation from others and atomic electrons are now under the influence of not only their own nucleus but also the other neighbouring atoms. We then have to consider the energy levels as pertaining to the solid as a whole, rather than an individual atom. The effect of this packing together of atoms into a solid is to give bands of energy levels, rather than discrete levels. The highest energy band containing occupied levels at a temperature of 0 K in a solid is called the *valence band*. The energy band immediately above the valence energy band and which contains vacant energy levels at 0 K is called the *conduction band*. The energy gap between the valence band and the conduction band depends on the element or compound concerned.

When a potential difference is applied across a piece of material, an electric field is produced within that material. In the case of a good conductor, this potential difference produces a current. The electric field thus gives rise to forces on charge carriers, i.e. electrons in the case of metals, and they are able to freely move. This can only occur if the electric field is able to move valence electrons to empty energy levels. Thus we can offer a model for a good conductor as having a valence band with no energy gap between it and the conduction band (Figure 12.1(a)). In the case of an insulator, the application of a potential difference produces no current. Thus although an electric field is produced and acts on electrons, they are not able to move. The model for this is of a valence band with an energy gap between it and the conduction band, the energy gap being too large for the electrons to acquire sufficient energy to jump across it (Figure 12.1(b)). Diamond, an insulator, has an energy gap of about 5 eV (the electron-volt (eV) is a unit of energy and is the energy gained by an electron in moving though a potential difference of 1 V, i.e. about 1.6×10^{-19} J). A semiconductor is a material with a small energy gap between its valence band and conduction band, typically about 1 eV or less (Figure 12.1(c)). This is small enough for some electrons in the valence band to have jumped across it at room temperature, by virtue of the energy acquired by the temperature having being raised from 0 K to room temperature. Thus there are electrons in the conduction band with free energy levels into which they can easily move. Also, in the valence band there are some empty slots, termed holes, from where the electrons have jumped into the conduction band. Thus electrons in the valence band do have some empty holes into which they can move.

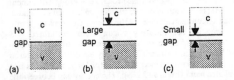

Figure 12.1 The energy bands for (a) a good conductor, (b) an insulator, (c) a semiconductor. c = conduction band, v = valence band

Semiconductors

The semiconductor elements, germanium and silicon, have atoms which bond together in the solid with each atom forming four covalent bonds with neighbouring atoms. Each such bond involves a shared pair of electrons. When a valence electron moves, one of the bonds is broken and so there is a hole left in the structure. Under the action of an electric field, produced by applying a potential difference across a piece of the material, an electron from a neighbouring atom can break its bond and move into the hole and so the hole moves to a new location. Thus conduction occurs as a result of the movement of electrons and holes. It is useful to think of the holes as positively charged particles since, under the action of an electric field, they move in the opposite direction to that of the electrons. Since all the holes are caused by the freeing of electrons from the bonds, i.e. from the valence band to the conduction band, there will be equal numbers of conduction electrons and holes. Such a semiconductor is said to be *intrinsic*.

The properties of semiconductors can be markedly changed by *doping*, i.e. the addition of small amounts of other elements. Silicon and germanium atoms have four valence electrons. If an element having five valence electrons, i.e. a group V element in the periodic table such as arsenic, antimony or phosphorus, is added so that some silicon atoms are replaced by such atoms then four of the five electrons settle into covalent bonds with neighbouring silicon atoms with the fifth electron easily donated for conduction. We can represent this situation by drawing an energy level, called the donor level, in the energy gap between the valence and conduction bands and close to the conduction band (Figure 12.2(a)). The energy gap between this donor level and the conduction band is of the order of 0.01 eV. Thus at room temperature, virtually all the donor electrons will have moved into the conduction band. The result is that there are more electrons in the conduction band than holes in the valence band. Electrical conduction is thus more by the movement of electrons than holes. Such a semiconductor is termed *n-type*, the n being because conduction is predominantly by negative charge carriers.

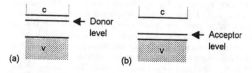

Figure 12.2 (a) n-type, (b) p-type

If an element having three valence electrons, i.e. a group III element from the Periodic table such as boron, aluminium, indium or gallium, is used to dope silicon then all three of the valence electrons settle into covalent bonds with neighbouring silicon atoms but there is a deficiency of one electron and so there is one bond which is incomplete. A hole has been introduced. The energy bands for the material then show an energy level, termed the acceptor level, with holes just above the valence band (Figure 12.2(b)). The energy gap between the acceptor level and the valence band is of the order of 0.01 eV. Electrons from the valence band can easily move into the acceptor level. There are now more holes in the valence band than electrons in the conduction band and so electrical conduction is predominantly by means of holes, hence the material is termed *p-type* with the p being because conduction is effectively by means of positive charge carriers. Doped semiconductors are called *extrinsic*.

Semiconductors are not restricted to just the group IV elements of silicon and germanium. Semiconductors have been produced by compounds between group III and group V elements, e.g. gallium arsenide, compounds between group II and group VI elements, e.g. cadmium sulphide, and between group IV and group VI elements, e.g. lead sulphide.

Dielectrics

Dielectric materials are insulators. Their properties determine the performance of insulation, e.g. that for cables, and capacitors. The *relative permittivity* er is an important property. For a parallel plate capacitor with plates of area A, separated by a distance d, the capacitance C is given by

$$C = \frac{\varepsilon_r \varepsilon_0 A}{d} = \varepsilon_r C_0$$

where ε_0 is the permittivity of free space and has a value of 8.85×10^{-12} F/m and $C_0 = \varepsilon_0 A/d$ and is the capacitance the capacitor would have with a vacuum between the plates. The relative permittivity is thus the factor by which the capacitance is multiplied by the presence of the dielectric between the capacitor plates.

When an alternating voltage is applied across a capacitor, there is an energy loss due to the reorientation of the atoms and molecules in the dielectric. This *dielectric loss* is expressed as the tangent of an angle, tan d. This expresses the ratio

$$\tan \delta = \frac{\text{energy lost per cycle}}{2\pi + \text{maximum energy stored}}$$

The smaller the dielectric loss, the less the oscillating charges within the dielectric dissipate energy as heat.

If the voltage across a capacitor is increased from zero, the current between the plates will be negligible until the breakdown voltage is reached. The current then rises rapidly. The electric field strength at which breakdown occurs is termed the *dielectric strength*. Since the electric field strength is the voltage gradient, the dielectric strength is the breakdown voltage divided by the thickness of the dielectric across which the voltage is applied.

12.2 Properties

Conductivity

Table 12.1 shows the resistivities and conductivities of a range of commonly used solid metals and alloys at about 20°C. Note that conductance is the reciprocal of resistance and has the unit siemen (S) and that conductivity is the reciprocal of the resistivity and has the unit S/m. In engineering, conductivity is often expressed as a percentage of the conductivity that annealed copper has at 20°C. Such values are said to be IACS values. Table 12.2 shows the resistances of wires of commonly used conductors. Table 12.3 shows resistivities for insulators (see also Table 9.7).

Table 12.1   Resistivities of conductors at 20°C

Material	Resistivity $(10^8\ \Omega)$	IACS value (%)
Aluminium (99.996% pure)	2.65	64.9
Brass, cartridge (70%)	6.2	28
yellow	6.4	27
Constantan (55% Cu, 45% Ni)	49.9	3.5
Copper (>99.90 %,electrolytic)	1.71	101
(>99.95%, oxygen free)	1.71	101
−1% cadmium wire	2.2	80
−15% zinc alloy	4.7	37
−20% zinc alloy	5.4	32
−2% nickel alloy	5.0	35
−6% nickel alloy	9.9	17
Gold	2.35	75
Iron (99.99% pure)	9.7	17.7
−0.65% C (carbon steel)	18	9.5
Manganin (87% Cu, 13% Mn)	48.2	3.5
Nichrome (80% Ni, 20% Cu)	108	1.6
Nickel (99.8%)	8.0	23
Phosphor bronze (3%)	8.6	20
Platinum	10.6	16
−10% iridium alloy	25	7
−10% rubidium alloy	43	4
Silver	1.59	106
−10% copper alloy	2	85
−15% cadmium alloy	4.9	35
Steel, stainless	56	3.1
17% cobalt	28	6.3
Tungsten	5.65	30

Table 12.2   Resistances of wires at 20°C

SWG	Diameter (mm)	Resistance in Ω/m			
		Copper	Manganin	Constantan	Nichrome
12	2.642	0.00312	0.076	0.090	0.197
14	2.032	0.00532	0.128	0.151	0.333
16	1.626	0.00831	0.200	0.235	0.520
18	1.219	0.0148	0.355	0.420	0.92
20	0.914	0.0263	0.630	0.745	1.65
22	0.711	0.0434	1.05	1.23	2.72
24	0.559	0.0703	1.69	2.00	4.40
26	0.457	0.105	2.53	3.00	6.60
28	0.376	0.155	3.75	4.40	9.70
30	0.315	0.221	5.30	6.30	13.9

Table 12.3 Resistivities of insulators at 20°C

Material	Resistivity (Ω/m)
Ceramic: alumina	10^9 to 10^{12}
porcelain	10^{10} to 10^{12}
Diamond 10^{10} to 10^{11}	
Glass: soda lime	10^9 to 10^{11}
Pyrex	10^{12}
Elastomer: Butyl	10^{15}
natural rubber	10^{13} to 10^{15}
polyurethane	10^{10}
Mica	10^{11} to 10^{15}
Paper (dry)	10^{10}
Polymer: acrylic	10^{12} to 10^{14}
cellulose acetate	10^8 to 10^{12}
melamine	10^{10}
polyamide (nylon)	10^{10} to 10^{13}
polypropylene	10^{13} to 10^{15}
polythene	
High density	10^{14} to 10^{15}
Low density	10^{14} to 10^{18}
polyvinylchloride	
Rigid	10^{12} to 10^{14}
Flexible	10^9 to 10^{13}

Band gap widths for semiconductors

Table 12.4 shows the energy gap widths between the valence and conduction bands for pure semiconductor elements and compounds.

Table 12.4 Band gap widths at 300 K

Material	Band gap width (eV)
Elements	
Silicon	1.12
Germanium	0.66
III-IV compounds	
Gallium arsenide GaAs	1.43
Gallium phosphide GaP	2.24
Gallium antimonide GaSb	0.72
Indium arsenide InAs	0.33
Indium phosphide InP	1.29
Indium antimonide InSb	0.17
Aluminium arsenide AlAs	2.16
Aluminium antimonide AlSb	1.58
II-VI compounds	
Cadmium sulphide CdS	2.42
Cadmium selenide CdSe	1.70
Zinc sulphide ZnS	3.68
Zinc selenide ZnSe	2.7
IV-VI compounds	
Lead sulphide PbS	0.41
Lead selenide PbSe	0.27
Lead telluride PbTe	0.31
Tin telluride SnTe	0.18

Dielectric properties

Table 12.5 shows typical values for relative permittivities, loss factors and dielectric strengths of commonly used dielectrics. The relative permittivity and loss factor are affected by frequency. The values quoted are those typical of a frequency of about 50/60 Hz.

Table 12.5   Properties of common dielectrics

Material	Relative permittivity	Loss factor	Dielectric strength (10^6 V/m)
Alumina	9 to 6.5	0.002	6
Acrylics	3.2	0.02	18
Cellulose acetate	3.5 to 7.5	0.01 to 0.1	12 to 24
Epoxy resin	3.6	0.02	18
Glass, soda lime	7·	0.01	10
Melamine	7	0.04	15
Mica	7	0.001	40
Paper	5	0.01	16
Polyamide,			
nylon 6/6	3.6 to 4.0	0.01	14
nylon 6·10	4.0 to 7.6	0.05	
Polycarbonate	2.8	0.0003	16
Polyethylene	2.3	0.0001	16
Polypropylene	2.1 to 2.7	0.005	18 to 26
Polystyrene,	2.5 to 2.7	0.0001	20 to 28
high impact	2.5 to 3.5	0.004	20
Polyvinylchloride,			
flexible		0.1	12 to 40
rigid	5 to 9	0.01	17 to 40
PTFE	2.4	0.0001	16
Rubber, natural	4 to 3.2	0.02	20
Vacuum	1	0	
Air (dry)	1	0	3
Water	80	Large ≈ 0.1	

13 Magnetic properties

13.1 *Terminology*

The *magnetic flux density B* is the amount of flux passing through a unit area, a measure of this being given by the density of magnetic field lines. The *magnetic field strength H* is the input to the system which is responsible for the flux. It is defined as being NI/L, where N is the number of turns of a coil carrying a current I and L the length of the flux path in which flux is produced by the coil. In a vacuum

$$B = \mu_0 H$$

where μ_0 is the permeability of free space and has the value 4p 10-7 H/m. When some other material is used for the flux path

$$B = \mu_r \mu_0 H$$

where μ_r is the relative permeability. The relative permeability thus specifies by what factor the flux density in a material is increased when compared with what it would have been in a vacuum. For most non-magnetic materials μ_r is independent of H and is about 1.

Diamagnetic materials have relative permeabilities slightly below 1, *paramagnetic* materials have relative permeabilities slightly greater than 1 and *ferromagnetic* and *ferrimagnetic* materials, the so-termed magnetizable materials, have relative permeabilities considerably greater than 1. Ferromagnetic materials are metals and ferrimagnetic materials are ceramics.

For magnetizable materials, the relative permeability for a particular material is not generally the same for all values of H. This means that a graph of B against H, i.e. the magnetization curve, is not a straight line but is generally like that shown in Figure 13.1 for cast steel.

Figure 13.1 *B-H* relationship for cast steel

Figure 13.2 shows how the flux density B in a magnetizable material changes as the magnetic field strength H is changed; starting with $H = 0$, increasing until B no longer increases and reaches a maximum value B_{max}. Then H is reduced back to 0, B however not reducing to 0 but some flux being retained. The retained flux density Br is termed the *remanence*. The direction of H is then reversed. When the flux density becomes 0 then the field strength necessary is termed the *coercive field* H_c. Further increase in H results in the flux density reversing in direction and again reaching a maximum. When H is then reduced to zero and then reversed in direction again, we have the *B-H* loop shown in Figure 13.2. This is termed the *hysteresis loop*. The area enclosed by the loop is the energy dissipated per unit volume in taking the material through one cycle of the magnetizing field.

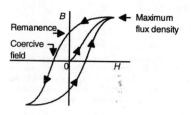

Figure 13.2 Hysteresis loop

The term *soft* when applied to magnetic materials is used for those which have a low coercive field so that only a small magnetic field is required to magnetize or demagnetize it, a small remanence so that only a little magnetic flux is retained in the absence of the magnetising field and a small area enclosed by the hysteresis loop so that little energy is lost per cycle of magnetization. Soft magnetic materials are used for items such as transformer cores and so subject to cycling. The term *hard* is used for magnetic materials having a high remanence so that a high amount of magnetic flux is retained in the absence of a magnetic field, a high coercive field so that it is difficult to demagnetize the material, and a large area enclosed by the hysteresis loop so that large amounts of energy are needed for demagnetization.

With soft magnetic materials it is usual to quote the maximum value of the relative permeability. This is the maximum gradient line that can be drawn to the magnetization curve. With hard magnetic materials for permanent magnets, the material is not usually cycled round the hysteresis loop by alternating magnetic fields. A quantity used to give a measure of the suitability of the material for a permanent magnet is the maximum value of the product BH.

13.2 Magnetic properties

Table 13.1, in two parts, gives the properties of some commonly encountered soft magnetic materials and Table 13.2 those for hard magnetic materials. The Curie temperature is the temperature at which thermal energy results in the loss of ferromagnetism.

Table 13.1 Soft magnetic materials

Material	Max. B (T)	Max. μ_r	Coercive field (A/m)	Energy loss/cycle (J/m³)
Pure iron	2.2	200 000	4	30
Mild steel	2.1	2 000	143	500
Silicon iron Fe + 3% Si	2.0	30 000	12	30
Permalloy Fe + 78.5% Ni	1.1	100 000	4	4
Supermalloy 79% Ni, 16% Fe, 5% Mo	0.8	800 000	0.16	4
Ferroxcube Mn Zn ferrite	0.25	1 500	0.8	13

Material	Curie temp. (K)	Resistivity ($\mu\Omega$m)
Pure iron	1043	0.1
Mild steel	1000	0.1
Silicon iron		
Fe + 3% Si	1030	0.5
Permalloy	800	0.2
Fe + 78.5% Ni		
Supermalloy	620	0.6
79% Ni, 16% Fe, 5% Mo		
Ferroxcube	570	10^6
Mn Zn ferrite		

Table 13.2 Hard magnetic materials

Material	Remanence (T)	Coercive field (kA/m)	Max. (BH) (kJ/m³)	Curie temp. (K)
Isotropic				
Alni				
Magloy 6	0.56	46	10	1030
Alnico				
Magloy 5	0.72	45	13.5	1070
*Feroba 1	0.22	135	8	720
*Ferroxdure 100				
*Neoperm D1				
Anisotropic				
Alcomax 3	1.26	52	43	1130
Magloy 1				
Ticonal 600				
Columax	1.35	59	60	1130
Magloy 100X				
*Feroba 2	0.39	150	29	720
*Ferroxdure 300				
*Neoperm E2				
*Feroba 3	0.37	240	26	720
*Ferroxdur 380				
*Neoperm E3				
Steels				
6% tungsten	1.05	5.2	2.4	1030
6% chromium	0.95	5.2	2.4	1030
3% cobalt	0.72	10	2.8	1070
15% cobalt	0.83	14	4.9	1110

A number of alternative trade names are given for entries.
*These are ferrites. The other isotropic and anisotropic materials are all iron, cobalt, nickel, aluminium alloys.

14 Mechanical properties

14.1 *Static strength*

Static strength can be defined as the ability to resist a short-term steady load at moderate temperatures without breaking or crushing or suffering excessive deformations. If a component is subject to a uniaxial stress the yield stress is commonly taken as a measure of the strength if the material is ductile and the tensile strength if it is brittle. Measures of static strength are thus yield strength, proof stress, tensile strength, compressive strength and hardness, the hardness of a material being related to the tensile strength of a material.

If the component is subject to biaxial or triaxial stresses, e.g. a shell subject to internal pressure, then there are a number of theories which can be used to predict material failure. The maximum principal stress theory, which tends to be used with brittle materials, predicts failure as occurring when the maximum principal stress reaches the tensile strength value, or the elastic limit stress value, that occurs for the material when subject to simple tension. The maximum shear stress theory, used with ductile materials, considers failure to occur when the maximum shear stress in the biaxial or triaxial stress situation reaches the value of the maximum shear stress that occurs for the material at the elastic limit in simple tension. With biaxial stress, this occurs when the difference between the two principal stresses is equal to the elastic limit stress. Another theory that is used with ductile materials is that failure occurs when the strain energy per unit volume is equal to the strain energy at the elastic limit in simple tension.

It should be recognized that a requirement for strength in a component requires not only a consideration of the static strength of the material but also the design. Thus for bending, an I-beam is more efficient than a rectangular cross-section beam because the material in the beam is concentrated at the top and bottom surfaces where the stresses are high and is not wasted in regions where the stresses are low. A thin shell or skin can be strengthened by adding ribs or corrugations.

For most ductile wrought materials, the mechanical properties in compression are sufficiently close to those in tension for the more readily available tensile properties to be used as an indicator of strength in both tension and compression. Metals in the cast condition, however, may be stronger in compression than in tension. Brittle materials, such as ceramics, are generally stronger in compression than in tension. There are some materials where there is significant anisotropy, i.e. the properties depend on the direction in which it is measured. This can occur with, for example, wrought materials where there are elongated inclusions and the processing results in them becoming orientated in the same direction or in composite materials containing unidirectional fibres.

The mechanical properties of metals are very much affected by the treatment they undergo, whether it be heat treatment or working. Thus it is not possible to give anything other than a crude comparison of alloys in terms of tensile strengths. The properties of polymeric materials are very much affected by the additives mixed in with them in their formulation and thus only a crude comparison of mechanical properties of different polymers is possible. There is also the problem with thermoplastics in that, even at 20°C they can show quite significant creep and this is more marked as the temperature increases. Thus their strengths are very much time-dependent. Unreinforced thermoplastics have low strengths when compared with most metals, however their low density means they have a favourable strength to weight ratio.

Table 14.1 gives a general comparison of tensile strengths of a range of materials, all the data referring to temperatures around about 20°C. Table 14.2 gives a general comparison of typical specific strengths, i.e. tensile or yield stress divided by the density to give a measure of strength per unit mass. Table 14.3 gives commonly used steels for different levels of tensile strength, the relevant limiting ruling sections being quoted.

Table 14.1 Materials according to strengths

Strength (MPa)	Material
<10	Polymer foams
2 to 12	Woods perpendicular to the grain
2 to 12	Elastomers
6 to 100	Woods parallel to the grain
60 to 100	Engineering polymers
20 to 60	Concrete
20 to 60	Lead alloys
80 to 300	Magnesium alloys
160 to 400	Zinc alloys
100 to 600	Aluminium alloys
80 to 1000	Copper alloys
250 to 1300	Carbon and low alloy steels
250 to 1500	Nickel alloys
500 to 1800	High alloy steels
100 to 1800	Engineering composites
1000 to >10 000	Engineering ceramics

Table 14.2 Specific strength at 20°C

Material	Density (Mg/m^3)	Strength to weight ratio (MPa/Mg m^{-3})
Aluminium alloys	2.6 to 2.9	40 to 220
Copper alloys	7.5 to 9.0	8 to 110
Lead alloys	8.9 to 11.3	1 to 3
Magnesium alloys	1.9	40 to 160
Nickel alloys	7.8 to 9.2	30 to 170
Titanium alloys	4.3 to 5.1	40 to 260
Zinc alloys	5.2 to 7.2	30 to 60
Carbon and low alloy steels	7.8	30 to 170
High alloy steels	7.8 to 8.1	60 to 220
Engineering ceramics	2.2 to 3.9	>300
Glasses	2 to 3	200 to 800
Thermoplastics	0.9 to 1.6	15 to 70
Polymer foams	0.04 to 0.7	0.4 to 12
Engineering composites	1.4 to 2	70 to 900
Concrete	2.4 to 2.5	8 to 30
Wood	0.4 to 1.8	5 to 60

Note: the unit used for density is Mg/m^3 which is 1000 kg/m^3.

Table 14.3 Steel selection

Tensile strength (MPa)	BS Steel code	Description of steel	Limit. ruling section (mm)
620 to 770	080M40	Medium carbon steel, hardened and tempered.	63
	150M36	Carbon-Mn steel, hardened and tempered.	150
	503M40	1% nickel steel, hardened and tempered.	250
700 to 850	150M36	1.5% manganese steel, hardened and tempered.	63
	708M40	1% Cr-Mo steel, hardened and tempered.	150
	605M36	1.5% Mn-Mo steel, hardened and tempered.	250
770 to 930	708M40	1% Cr-Mo steel, hardened and tempered.	100
	817M40	1.5% Ni-Cr-Mo steel, hardened and tempered.	250
850 to 1000	630M40	1% Cr steel, hardened and tempered.	63
	709M40	1% Cr-Mo steel, hardened and tempered.	100
	817M40	1.5% Ni-Cr-Mo steel, hardened and tempered.	250
930 to 1080	709M40	1.5% Cr-Mo steel, hardened and tempered.	63
	817M40	1.5% Ni-Cr-Mo steel, hardened and tempered.	100
	826M31	2.5% Ni-Cr-Mo steel, hardened and tempered.	250
1000 to 1150	817M40	1% Ni-Cr-Mo steel, hardened and tempered.	63
	826M31	2.5% Ni-Cr-Mo steel, hardened and tempered.	150
1080 to 1240	826M31	2.5% Ni-Cr-Mo steel, hardened and tempered.	100
	826M40	2.5% Ni-Cr-Mo steel, hardened and tempered.	250
1150 to 1300	826M40	2.5% Ni-Cr-Mo steel, hardened and tempered.	150
1240 to 1400	826M40	2.5% Ni-Cr-Mo steel, hardened and tempered.	150
>1540	835M30	4% Ni-Cr-Mo steel, hardened and tempered.	150

14.2 Stiffness

Stiffness can be considered to be the ability of a material to resist deflection when loaded. Thus if we consider a cantilever of length L subject to a point load F at its free end, then the deflection y at the free end is given by

$$y = \frac{Fl^3}{3EI}$$

with E being the tensile modulus and I the second moment of area of the beam cross-section with respect to the neutral axis. Thus for a given shape and length cantilever, the greater the tensile modulus the smaller the deflection. Similar relationships exist for other forms of beam. Hence we can state that the greater the tensile modulus the greater the stiffness.

The deflection of a beam is a function of both E and I. Thus, for a given material, a beam can be made stiffer by increasing its second moment of area. The second moment of area of a section is increased by placing as much as possible of the material as far as possible from the axis of bending. Thus an I-section is a particularly efficient way of achieving stiffness. Similarly a tube is more efficient than a solid rod.

Another situation which is related to the value of EI is the *buckling* of columns when subject to compressive loads. The standard equation used for buckling has it occurring for a column of length L when the load F reaches the value

$$F = \frac{\pi^2 EI}{L^2}$$

This is Euler's equation. The bigger the value of EI the higher the load required to cause buckling. Hence we can say that the column is stiffer the higher the value of EI. Note that a short and stubby column is more likely to fail by crushing when the yield stress is exceeded rather than buckling. Buckling is however more likely to be the failure mode if the column is slender.

The tensile modulus of a metal is little affected by changes in its composition or heat treatment. However, the tensile modulus of composite materials is very much affected by changes in the orientation of the fillers and the relative amounts. Table 14.4 shows typical tensile modulus values for materials at 20°C.

Table 14.4 Materials according to tensile modulus

Tensile modulus (GPa)	Material
<0.2	Polymer foams
<0.2	Elastomers
0.2 to 10	Woods parallel to grain
0.2 to 10	Engineering polymers
2 to 20	Woods perpendicular to grain
10 to 11	Lead alloys
20 to 50	Concrete
40 to 45	Magnesium alloys
50 to 80	Glasses
70 to 80	Aluminium alloys
43 to 96	Zinc alloys
110 to 125	Titanium alloys
100 to 160	Copper alloys
200 to 210	Steels
80 to 1000	Engineering ceramics

14.3 Fatigue resistance

The failure of a component when subject to fluctuating loads is as a result of cracks which tend to start at some discontinuity in the material and grow until failure occurs. The main factors affecting fatigue properties are stress concentrations caused by component design, corrosion, residual stresses, surface finish/treatment, temperature, the microstructure of the alloy and its heat treatment. Only to a limited extent does the choice of material determine the fatigue resistance of a component.

In general, for metals the fatigue limit or endurance limit at about 10^7 to 10^8 cycles lies between about a third and a half of the static tensile strength. For steels the fatigue limit is typically between 0.4 and 0.5 that of the static strength. Inclusions in the steel, such as sulphur or lead to improve machinability, can, however, reduce the fatigue limit. For grey cast iron the fatigue limit is about 0.4 that of the static strength, for nodular and malleable irons in the range 0.5 for ferritic grades to 0.3 for the higher strength pearlitic irons, for blackheart, whiteheart and the lower strength pearlitic malleable irons about 0.4. With aluminium alloys the endurance limit is about 0.3 to 0.4 that of the static strength, for copper alloys about 0.4 to 0.5.

Fatigue effects with polymers are complicated by the fact that the alternating loading results in the polymer becoming heated. This causes the elastic modulus to decrease and at high enough frequencies this may be to such an extent that failure occurs. Thus fatigue in polymers is very much frequency dependent.

14.4 Toughness

Toughness can be defined as the resistance offered by a material to fracture. A tough material is resistant to crack propagation. A measure of toughness is given by two main measurements: the resistance of a material to impact loading which is measured in the Charpy or Izod tests by the amount of energy needed to fracture a test piece and the resistance of a material to the propagation of an existing crack in a fracture toughness test. This yields a parameter termed the plain strain fracture toughness K_{Ic}, the lower its value the less tough the material. Table 14.5 gives typical values of the plane strain fracture toughness at 20°C.

Within a given type of metal alloy there is an inverse relationship between yield stress and toughness, the higher the yield stress the lower the toughness. Thus if, for instance, the yield strength of low alloy, quenched and tempered steels is pushed up by metallurgical means then the toughness declines. Steels become less tough with increasing carbon content and larger grain size.

The toughness of plastics is improved by incorporating rubber or another tougher polymer, copolymerization, or incorporating tough fibres. For example, styrene-acrylonitrile (SAN) is brittle and far from tough. It can however be toughened with the rubber polybutadiene to give the tougher acrylonitrile-butadiene-styrene (ABS).

Table 14.5 Plane strain fracture toughness at 20°C

Plain strain fracture toughness (MN m$^{-3/2}$)	Material
<1.0	Polymer foams
0.07 to 0.9	Woods perpendicular to grain
0.1 to 0.3	Concrete
0.3 to 0.6	Glasses
0.5 to 10	Engineering polymers
1 to 10	Woods parallel to the grain
2 to 10	Engineering ceramics
7 to 11	Cast irons
10 to 11	Magnesium alloys
10 to 60	Aluminium alloys
10 to 100	Engineering composites
20 to 150	Steels
50 to 110	Copper alloys
60 to 110	Titanium alloys
60 to 110	Nickel alloys

14.5 Creep and temperature resistance

The creep resistance of a metal can be improved by incorporating a fine dispersion of particles to impede the movement of dislocations. The Nimonic series of alloys, based on an 80/20 nickel-chromium alloy, have good creep resistance as a consequence of fine precipitates formed by the inclusion of small amounts of titanium, aluminium, carbon or other elements. Creep increases as the temperature increases and is thus a major factor in determining the temperature at which materials can be used. Another factor is due to the effect on the material of the surrounding atmosphere. This can result in surface attack and scaling which gradually reduces the cross-sectional area of the component and so its ability to carry loads. Such effects increase as the temperature increases. The Nimonic series of alloys have good resistance to such attack. Typically they can be used up to temperatures of the order of 900°C.

For most metals creep is essentially a high temperature effect, however this is not the case with plastics. Here creep can be significant at room temperatures. Generally thermosets have higher temperature resistance than thermoplastics, however the addition of suitable fillers and fibres can improve the temperature properties of thermoplastics. Table 14.6 indicates typical temperature limitations for a range of materials.

Table 14.6 Temperature limitations of materials

Temperature limit (°C)	Materials
Room temp. to 150	Few thermoplastics are recommended for prolonged use above about 100°C. Glass-filled nylon can however be used up to 150°C. The only engineering metal which has limits within this range is lead.
150 to 400	Magnesium and aluminium alloys can in

Table 14.6 (continued)

Temperature limit (°C)	Materials
	general only be used up to about 200°C, though some specific alloys can be used to higher temperatures. For example, the aluminium alloy LM13 (AA336.0) is used for pistons in engines and experiences temperatures of the order of 200 to 250°C while some cast aluminium bronzes can be used up to about 400°C with wrought aluminium bronzes up to about 300°C. Plain carbon and manganese-carbon steels are widely used for temperatures in this range.
400 to 600	Plain carbon and manganese-carbon steels cannot be used above about 400–450°C. For such temperatures low-alloy steels are used. For temperatures up to about 500°C a carbon–0.5% Mo steel might be used, up to about 525°C a 1% Cr–0.5% Mo steel, and up to about 550°C a 0.5% Cr-Mo-V steel, and up to about 600°C a steel with 5 to 12% Cr. Titanium alloys are also widely used in this temperature range. The alpha-beta alloy 6% Al–4% V (IMI318) is used up to about 450°C. Near alpha alloys can be used to higher temperatures, for example the alloy IMI 829 is used up to about 600°C.
600 to 1000	Metals most widely used in this temperature range are the austenitic stainless steels, Ni-Cr and Ni-Cr-Fe alloys, and cobalt base alloys. Austenitic stainless steels with 18% Cr–8% Ni can be used up to about 750°C. A range of high temperature alloys based on the nickel-chromium base are able to maintain their strength, resistance to creep and oxidation resistance at high temperatures, e.g. Nimonic series alloys such as Nimonic 90 which can be used up to about 900°C, Nimonic 901 to about 1000o C. Another series of high temperature alloys are the Ni-Cr-Fe alloys, such as the Inconel and Incoloy series. For example, Inconel 600 can be used up to virtually 1000°C and Incoloy 800H to 700°C. Above 1000 The materials which can be used at temperatures in excess of 1000°C are the refractory metals, i.e. molybdenum, niobium, tantalum and tungsten, and ceramics. The refractory metals, and their alloys, can be used at temperatures in excess of 1500°C. Surface protection is one of the main problems facing the use of these alloys at high temperatures. Ceramics can also be used at such high temperatures but tend to suffer from the problems of being hard, brittle and

Table 14.6 Temperature limitations of materials

Temperature limit (°C)	Materials
	vulnerable to thermal shock. Alumina is used in furnaces up to about 1600°C, silicon nitride to about 1200°C and silicon carbide to about 1500°C.

14.6 Selection criteria

Strength to density and elastic modulus to density ratios are both used as general performance indices for materials, the intention being to maximize such ratios in order to obtain the best performance for the least mass. These may not give the best performance in all loading situations. Table 14.7 give the criteria for maximizing the strength to mass and stiffness to mass ratios for a range of different loading conditions where failure is considered to be due to excessive deflection.

Table 14.7 Performance indices

Component	Maximize stiffness	Maximize strength
Tie, i.e. tensile strut	$\dfrac{E}{\rho}$	$\dfrac{\sigma_y}{\rho}$
Beam	$\dfrac{E^{1/2}}{\rho}$	$\dfrac{\sigma_y^{2/3}}{\rho}$
Column, i.e. compressive strut	$\dfrac{E^{1/2}}{\rho}$	$\dfrac{\sigma_y}{\rho}$
Plate, loaded externally or by self weight in bending	$\dfrac{E^{1/3}}{\rho}$	$\dfrac{\sigma_y}{\rho}$
Cylinder with internal pressure	$\dfrac{E}{\rho}$	$\dfrac{\sigma_y}{\rho}$
Sperical shell with internal pressure	$\dfrac{E}{(1-v)\rho}$	$\dfrac{\sigma_y}{\rho}$

Note: E is the modulus of elasticity, ρ the density, σ_y the yield stress (though sometimes the tensile strength is used), v Poissons ratio.

15 Corrosion and wear resistance

15.1 *Corrosion resistance*

For metals subject to atmospheric corrosion the most significant factor in determining the chance of corrosive attack is whether there is an aqueous electrolyte present. This could be provided by condensation of moisture occurring as a result of the climatic conditions. The amount of pollution in the atmosphere can also affect the corrosion rate. Corrosion can often be much reduced by the selection of appropriate materials. For metals immersed in water, the corrosion depends on the substances that are dissolved or suspended in the water.

Carbon steels and low alloy steels are not particularly corrosion resistant, rust being the evidence of such corrosion. In an industrial atmosphere, in fresh and sea water, plain carbon steels and low alloy steels have poor resistance. Painting, by providing a protective coating of the surface, can reduce such corrosion. The addition of chromium to steel can markedly improve its corrosion resistance. Steels with 4–6% chromium have good resistance in an industrial atmosphere, in fresh and sea water, while stainless steels have an excellent resistance in an industrial atmosphere and fresh water but can suffer some corrosion in sea water. The corrosion resistance of grey cast iron is good in an industrial atmosphere but not so good in fresh or sea water, though still better than that of plain carbon steels.

Aluminium when exposed to air develops an oxide layer on its surface which then protects the substrate from further attack. Wrought alloys are often clad with thin sheets of pure aluminium or an aluminium alloy to enhance the corrosion resistance of such alloys. Thus in air, aluminium and its alloys have good corrosion resistance. When immersed in fresh or sea water, most aluminium alloys offer good corrosion resistance, though there are some exceptions which must be clad in order to have good corrosion resistance.

Copper in air forms a protective green layer which protects it from further attack and thus gives good corrosion resistance. Copper has also good corrosion resistance in fresh and sea water, hence the widespread use of copper piping for water distribution systems and central heating systems. Copper alloys likewise have good corrosion resistance in industrial atmospheres, fresh and sea water though demetallification can occur with some alloys, e.g. dezincification of brass with more than 15% zinc.

Nickel and its alloys have excellent resistance to corrosion in industrial air, fresh and sea water.

Titanium and its alloys have excellent resistance, probably the best resistance of all metals, in industrial air, fresh and sea water and are thus widely used where corrosion could be a problem.

Plastics do not corrode in the same way as metals and thus, in general, have excellent corrosion resistance. Hence, for example, the increasing use of plastic pipes for the transmission of water and other chemicals. Polymers can deteriorate as a result of exposure to ultraviolet radiation, e.g. that in the rays from the sun, heat and mechanical stress. To reduce such effects, specific additives are used as fillers in the formulation of a plastic.

Most ceramic materials show excellent corrosion resistance. Glasses are exceedingly stable and resistant to attack, hence the widespread use of glass containers. Enamels, made of silicate and borosilicate glasses, are widely used as coatings to protect steels and cast irons from corrosive attack.

Table 15.1 gives a rough indication of the corrosion resistance of materials to different environments.

Table 15.1 Corrosion resistance in various environments

Corrosion resistance	Material
Aerated water	
High resistance	All ceramics
	Glasses
	Lead alloys
	Alloy steels
	Titanium alloys
	Nickel alloys
	Copper alloys
	PTFE, polypropylene, nylon, epoxies, polystyrene, PVC
Medium resistance	Aluminium alloys
	Polythene, polyesters
Low resistance	Carbon steels
Salt water	
High resistance	All ceramics
	Glasses
	Lead alloys
	Stainless steels
	Titanium alloys
	Nickel alloys
	Copper alloys
	PTFE, polypropylene, nylon, epoxies, polystyrene, PVC, polythene
Medium resistance	Aluminium alloys
	Polyesters
Low resistance	Low alloy steels
	Carbon steels
UV radiation	
High resistance	All ceramics
	Glasses
	All alloys
Medium resistance	Epoxies, polyesters, polypropylene, polystyrene, HD polyethylene, polymers with UV inhibitor
Low resistance	Nylon, PVC, many elastomers
Strong acids	
High resistance	Glasses
	Alumina, silicon carbide, silica PTFE, PVC, polythene, epoxies, elastomers
	Lead alloys

Table 15.1 (continued)

Corrosion resistance	Material
	Titanium alloys
	Nickel alloys
	Stainless steels
Medium resistance	Magnesium oxide
Low resistance	Aluminium alloys
	Carbon steels
	Polystyrene, polyurethane, nylon, polyesters
Strong alkalis	
High resistance	Alumina
	Nickel alloys
	Steels
	Titanium alloys
	Nylon, polythene, polystyrene, PTFE, PVC, polypropylene, epoxies
Medium resistance	Silicon carbide
	Copper alloys
	Zinc alloys
	Elastomers, polyesters
Low resistance	Glasses
	Aluminium alloys
Organic solvents	
High resistance	All ceramics
	Glasses
	All alloys
	PTFE, polypropylene
Medium resistance	Polythene, nylon, epoxies
Low resistance	Polystyrene, PVC, polyesters, ABS, most elastomers

Dissimilar metal corrosion

Table 15.2 shows the galvanic series of metals in sea water. The series will differ if the environment is freshwater or industrial atmosphere, though the same rough sequence tends to occur but the potentials are likely to vary. The list is in order of corrosion tendency, giving the free corrosion potentials, and enables the prediction of the corrosion resistance of a combination of dissimilar metals. The bigger the separation of any two metals in the series, the more severe the corrosion of the more active of them when a junction between the pair of them is exposed to sea water. The more negative potential metal acts as the anode and the less negative or positive as the cathode in an electrochemical cell.

Table 15.3 shows the susceptibility of alloys to pitting, stress corrosion and demetallification. With pitting the corrosion results in the production of small holes in the metal surface. Stress corrosion is where corrosion occurs in certain environments when the alloy is subject to stress. Demetallification is galvanic corrosion occurring between the constituents in an alloy, resulting in the loss of a particular alloy constituent and a consequential reduction in strength of the alloy.

Table 15.2 Galvanic series

Metal	Free corrosion potential (V)
Magnesium	−1.60 to −1.63
Zinc	−0.9 to −1.2
Aluminium alloys	−0.75 to −1.0
Mild steel	−0.6 to −0.7
Low alloy steel	−0.58 to −0.62
Aluminium bronze	−0.3 to −0.4
Yellow/red brass	−0.3 to −0.4
Tin	−0.3 to −0.34
Copper	−0.3 to −0.37
Lead/tin solder (50/50)	−0.3 to −0.35
Aluminium brass	−0.27 to −0.34
Manganese bronze	−0.27 to −0.32
Silicon bronze	−0.25 to −0.29
Tin bronze	−0.24 to −0.31
Stainless steel (410, 416)	−0.25 to −0.35
Nickel silver	−0.22 to −0.27
Cupronickel (80/20)	−0.2 to −0.3
Stainless steel (430)	−0.2 to −0.25
Lead	−0.2 to −0.22
Cupronickel (70/30)	−0.18 to −0.22
Nickel–chromium alloys	−0.13 to −0.17
Siver braze alloys	−0.1 to −0.2
Nickel	−0.1 to −0.2
Silver	−0.1 to −0.15
Stainless steel (302, 304, 321, 347)	−0.05 to −0.1
Nickel–copper alloys	−0.02 to −0.12
Stainless steel (317, 317)	0 to −0.1
Titanium	+0.05 to −0.05
Platinum	+0.25 to +0.20
Graphite	+0.2 to +0.3

Table 15.3 Corrosive susceptibilities of alloys

Alloy	Pitting	Stress corrosion	Demetallification
Magnesium alloys	S	S	No
Aluminium alloys	S	S	No
Steels	S	S	S
Nickel alloys	SA	S	No
Stainless steels	SA	S	No
Copper alloys	No	S	S
Titanium alloys	SA	SA	No

Note: S = susceptible, SA = susceptible only in aggressive or severe conditions, No = not generally susceptible.

15.2 *Selection for wear resistance*

Wear is the progressive loss of material from surfaces as a result of contact with other surfaces. It can occur as a result of sliding or rolling contact between surfaces or from the movement of fluids containing particles over surfaces. Because wear is a surface effect, surface treatments and coatings play an important role in improving wear resistance. Lubrication can be considered to be a way of keeping surfaces apart and so reducing wear.

Mild steels have poor wear resistance. However, increasing the carbon content increases the wear resistance. Surface hardenable carbon or low alloy steels enable wear resistance to be improved as a result of surface treatments such as carburizing, cynaniding or carbonitriding. Even better wear resistance is provided by nitriding medium carbon chromium or chromium-aluminium steels, or by surface hardening high carbon high chromium steels. Grey cast iron has good wear resistance for many applications. Better wear resistance is, however, provided by white irons. Among non-ferrous alloys, beryllium coppers and cobalt-base alloys, such as Stellite, offer particularly good wear resistance.

Metallic materials for use as bearing surfaces need to be hard and wear resistant, with a low coefficient of friction, but at the same time sufficiently tough. Generally these requirements are met by the use of a soft, but tough, alloy in which hard particles are embedded. The soft alloy is capable of yielding to accommodate any localized high pressures resulting from slight misalignments and starting up, while the hard particles provide the wear resistance. Such materials include the white bearing metals, copper base bearing metals and aluminium base bearing metals. The white bearing metals are tin or lead based materials. The tin-base materials, known as Babbit metals, are tin–antimony–copper alloys with possibly some lead. The hard particles are provided by antimony–tin and copper–tin compounds. The lead based materials are lead–antimony–tin alloys with antimony-tin compounds providing the hard particles. The main copper based metals used are phosphor bronzes with 10–15% tin and copper–lead alloys. The main aluminium bearing materials are aluminium–tin alloys.

Self-lubricating plastics, e.g. nylon 6.6 with the lubricating additive of 18% PTFE-2% silicone, offer very good wear resistance and are widely used for low wear applications such as bearings and gears.

Table 15.4, in three parts, gives the properties generally relevant in selecting bearing materials for commonly used bearing materials.

Table 15.4 Properties of bearing materials

Material	Brinell hardness	Yield stress (MPa)	Strength (MPa)	Elastic modulus (GPa)
White metal: Tin-base	17 to 25	30 to 65	70 to 120	51 to 53
White metal: Lead-base	15 to 20	20 to 60	40 to 110	29
Copper-lead alloys	20 to 40	40 to 60	50 to 90	75
Phosphor bronze	70 to 150	130 to 230	280 to 420	80 to 95
Leaded tin bronze	50 to 80	80 to 150	160 to 30	95
Aluminium–tin alloys	70 to 75	50 to 90	140 to 210	73
Polymers	5 to 20		20 to 80	1 to 10

Material	Density (Mg/m³)	Fatigue strength (MPa)	Thermal conductivity (W/m K)
White metal: Tin-base	7.3 to 7.7	25 to 35	50
White metal: Lead-base	9.6 to 10	22 to 30	24
Copper-lead alloys	9.3 to 9.5	40 to 50	42
Phosphor bronze	8.8	90 to 120	42
Leaded tin bronze	8.8	80	42
Aluminium–tin alloys	2.9	130 to 170	160
Polymers	1.0 to 1.3	5 to 40	0.1

Table 15.4 (continued)

Material	Relative corrosion resistance	Relative wear resistance	Relative cost
White metal: Tin-base	5	2	7
White metal: Lead-base	4	3	1
Copper-lead alloys	3	5	1.5
Phosphor bronze	2	5	2
Leaded tin bronze	2	3	2
Aluminium–tin alloys	3	2	1.5
Polymers	5	5	0.3

Note: for relative corrosion and relative wear, the higher the rating the less the corrosion and the less the wear. For relative cost, the higher the rating the more the material costs.

16 Thermal properties

16.1 Terms

The *heat capacity* is the energy required to raise the temperature of a particular body of a material by one degree. The *specific heat capacity c* is the energy required to raise the temperature of 1 kg of a material by one degree. Thus if energy Q raises the temperature of a mass m by T K then

$$c = \frac{Q}{mT} \quad \text{J kg}^{-1}\text{ K}^{-1}$$

The *linear coefficient of thermal expansion a* of a solid is the change in length per unit length per degree change in temperature. This if the length changes from L_0 to L_t when the temperature changes by T degrees, then

$$a = \frac{L_t - L_0}{L_0 T} \quad \text{K}^{-1}$$

Polymers have large values of coefficient, roughly ten times those of metals and almost a hundred times those of ceramics.

The *thermal conductivity λ* of a material is a measure of the rate at which heat is transferred through a material. When heat Q is transferred per second across a plane of area A of a material and the temperature gradient is $\Delta T/\Delta x$ (Figure 16.1) then

$$\frac{Q}{A} = \lambda \frac{\Delta T}{\Delta x}$$

The thermal conductivity has the unit W m^{-1} K^{-1}.

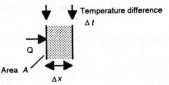

Figure 16.1 Thermal conduction

In the case of composite bodies, e.g. a cavity wall or a hot water tank consisting of a single metal sheet but with boundary layers of stagnant air and stagnant water, the term *overall heat transfer coefficient U* , or *U-value*, is often used.

$$\frac{Q}{A} = U\Delta T$$

Thus, for a single layer material we have $U = \lambda/\Delta x$.

16.2 Thermal properties

Table 16.1 gives values for the specific heat capacity, the coefficient of linear expansion and the thermal conductivity for a range of commonly used engineering materials at about 20°C. See also table 9.14 for more polymers, 10.5 and 10.6 for ceramics .

Table 16.1 Thermal properties

Material	α ($10^6 K^{-1}$)	c ($kJ\ kg^{-1}\ K^{-1}$)	λ ($W\ m^{-1}\ K^{-1}$)
Metals			
Aluminium	24	0.90	220 to 230
alloys	20 to 24	0.84	120 to 200
Copper	17	0.39	370
alloys	16 to 20	0.39	30 to 160
Iron	12	0.44	81
carbon steels	10 to 15	0.48	47
cast irons	10 to 11	0.27 to 0.46	44 to 53
alloy steels	12	0.51	13 to 48
stainless steel	11 to 16	0.51	16 to 26
Magnesium	25	1.02	156
alloys	25 to 27		80 to 140
Nickel	13	0.44	92
alloys	10 to 19	0.48 to 0.50	11 to 30
Tin	23	0.23	67
alloys	22 to 24		53 to 67
Titanium	8	0.52	22
alloys	8 to 9	0.54	5 to 12
Zinc	40	0.39	116
alloys	25 to 35		107 to 116
Polymers			
Thermoplastics	40 to 300	0.8 to 2.0	0.1 to 0.4
ABS	80 to 100	1.5	0.13 to 0.20
nylon	6 80 to 100	1.6	0.17 to 0.21
polythene	110 to 200	1.9 to 2.3	0.25 to 0.35
polypropylene	100 to 120	1.9	0.16
polystyrene	60 to 80	1.2	0.12 to 0.13
PVC	50 to 250	1.1 to 1.7	0.12 to 0.15
Thermosets	10 to 60	1.0 to 2.0	0.1 to 0.4
epoxy	60	1.1	0.17
phenol			
formaldehyde	30 to 40	1.6 to 1.8	0.13 to 0.25
Elastomers	50 to 250	1.3 to 1.8	0.1 to 0.3
natural rubber	22	1.9	0.18
neoprene	24	1.7	0.21
Cellular polymers			0.02 to 0.04
Ceramics			
Alumina	8 to 9	0.7	20 to 40
Bonded carbides	4 to 6	0.2 to 1.0	40 to 120
Glasses	3 to 9	0.5 to 0.7	0.5 to 2
Composites			
Engineering	7 to 20		0.3 to 2
Concrete	7 to 14	3.3	0.1 to 2
Wood		1.7	0.1 to 0.2
across grain	35 to 60		
along grain	3 to 6		
Glass fibre blanket			0.03 to 0.07

Table 16.2 gives some U-values for typical wall structures under normal conditions, i.e. not particularly sheltered or exposed to strong winds.

Table 16.2 U-value

Structure	Thickness (mm)	U-value (W m^{-2} K^{-1})
Solid brick, unplastered	105	3.3
	220	2.3
Solid brick with 16 mm plaster on inside	105	2.5 to 3.0
	220	1.9 to 2.1
Cavity wall with 105 mm brick and 16 mm plaster on inside	260	1.3 to 1.5
Cavity wall with 105 mm brick outer, 100 mm lightweight concrete inner and 16 mm plaster on inside	260	0.96
Cast concrete	150	3.5
	200	3.1
Tiled roof with roof space and 10 mm plasterboard ceiling		1.5
As above but with 50 mm glass fibre insulation between joists		0.5

17 Selection of materials

17.1 *The required specification*

A number of questions need to be answered before a decision can be made as to the specification required of a material and hence a decision made as to the optimum material for the task. The questions can be grouped under four general headings:

1 What properties are required?
2 What are the processing requirements and their implications for the choice of material?
3 What is the availability of materials?
4 What is the cost?

The following indicate the type of questions that are likely to be considered in trying to arrive at answers to the above general questions.

Properties

1 What mechanical properties are required? This means consideration of such properties as strength, stiffness, hardness, ductility, toughness, fatigue resistance, wear properties, etc. Coupled with the question is another one: Will the properties be required at low temperatures, about room temperature or high temperatures?
2 What electrical properties are required? For example, does the material need to be a good conductor of electricity or perhaps an insulator?
3 What magnetic properties are required? Does the material need to have soft or hard magnetic properties or perhaps be essentially non-magnetic?
4 What thermal properties are required? This means consideration of such properties as specific heat capacity, linear coefficient of expansion and thermal conductivity.
5 What chemical properties are required? This means considering the environment to which the material will be exposed and the possibility of corrosion.
6 What dimensional conditions are required? For example, does the material need to be capable of a good surface finish, have dimensional stability, be flat, have a particular size, etc.

Processing parameters

1 Are there any special processing requirements which will limit the choice of material? For example, does the material have to be cast or perhaps extruded?
2 Are there any material treatment requirements? For example, does the material have to be annealed or perhaps solution hardened?
3 Are there any special tooling requirements? For example, does the hardness required of a material mean special cutting tools are required?

Availability

1 Is the material readily available?
2 Are there any ordering problems for that material?
3 What form is the material usually supplied in? For example, is the material usually supplied in bars or perhaps sheet? This can affect the processes that can be used.

Cost

1 What is the cost of the raw material?
2 What are the cost implications of the process requirements?
3 What will be the cost of the processed material?

17.2 Selection criteria

In considering the materials likely to meet some specification, materials need to be considered in relation to how well they meet the specification. From the whole range of materials that exist, there is likely to be a subset which looks feasible and the issue is then of determining which give the optimum performance against the required specification. A number of methods have been used to determine the optimum material.

1 Identification of critical properties

The critical properties are those without which a material could not possibly be suitable. Thus there might, for example, be a requirement that the strength is above some particular stress limit, or that the electrical resistivity is below some particular value limit. By considering whether there are any such critical limits, a subset of materials can be arrived at.

One way of carrying out such an operation is to use charts with the materials indicated against their properties. Thus there might be a chart of the form shown in Figure 17.1 where the materials are indicated against both their values of elastic modulus and their density. To select materials which have at least a tensile modulus of, say, 100 GPa a line is drawn across the chart at that value and all the materials above that line form the subset from which a material can be selected. If we also have the requirement that the density must be less than 2 Mg/m3 then we draw a line on the chart at that value and all the materials to the left of that line form the subset with that criteria. The subset of materials with both criteria are thus those in the upper left quadrant.

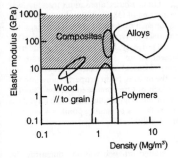

Figure 17.1 Property chart

Such a chart can also be used if a material is required in the form of a beam with the stiffness maximized (see Table 14.7). This requires the determination of the materials with values of E/r greater than some critical value. Suppose, for example, the subset of materials is required for which E/r is greater than 1000. This will be a line on the chart with a constant slope of 1000. Figure 17.2 shows such a line and the subset thus indicated.

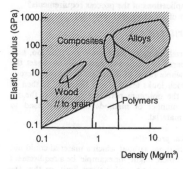

Figure 17.2 Property chart

A range of such charts can be found in *Materials Selection in Mechanical Design* (Pergamon Press 1992) by M.F.Ashby.

2 Merit rating

The problem can arise of trying to determine the optimum material when there are a number of required properties and a number of materials meeting the different properties in different ways. The issue is then to determine the material which achieves the best balance of properties. The merit rating method involves each material being given a relative merit value for each of the relevant properties. The ratings are assigned relative to some best material which is given the top rating value, say 10 or 100. All the others are then given a relative rating. For example, in considering electrical conductivity, pure copper (conductivity $59 \times 10-10$ S/m) may be given a rating of 100 and then other materials rated according to their conductivity as a percentage of that of copper. On this basis, aluminium (conductivity $38 \ 10-10$ S/m) might have a rating of $(38/59) \ 100 = 64$. When merit ratings have been allocated for each material and each property, the overall merit rating for a material is arrived at by taking the total of the weighted merit ratings for each property. For example, in considering the materials for the filament of an electric light bulb and the properties of melting point, electrical conductivity, strength and ductility we might consider the melting point property more important than the conductivity, which in turn is more important than the strength which is then in turn more important than ductility. As a result we might have weighting factors of melting point $\times 4$, conductivity $\times 3$, strength $\times 2$ and ductility $\times 1$.

3 Cost per unit property

Since low cost is often a requirement, one way of comparing the properties of a subset of materials is on the basis of cost per unit property or group of properties. Table 17.1 gives some relative costings of materials.

17.3 Relative costs

The costs of materials change with time, and to eliminate the need to specify any particular unit of currency, relative costs are used for the purpose of materials selection when we only require to determine the optimum material. The relative costs tend to be for mass defined in relationship to that of mild steel, often mild steel bar. Thus:

$$\text{relative cost} = \frac{\text{cost per kg of materials}}{\text{cost per kg of mild steel}}$$

Table 17.1 gives some typical values. The relative cost per kg can be converted into relative cost per m^3 by multiplying it by the density.

Table 17.1

Material	Relative cost per kg
Metals	
Mild steel bar (black)	1.0
Mild steel bar (bright)	1.3
Mild steel sheet	1.4
Medium carbon steel	1.6
High carbon steel bar	2.3
Cast iron casting	2.4
Manganese steel bar	2.5
Brass sheet	5.1
Copper sheet	8.3
Stainless steel sheet	8.5
Aluminium bar	8.5
Nickel chrome steel bar	4.6
Brass bar	6.6
Aluminium sheet	7.1
Aluminium casting	9.6
Stainless steel bar	9.6
Phosphor bronze bar	16.0
Monel bar	20.6
Polymers	
Polyethylene	3
Polypropylene	3
Polystyrene	3
PVC	6
ABS	12
Phenolics	12
Acrylics	12
Cellulose acetate	15
Acetals	15
Polycarbonate	36
Nylons	45
Polyurethane	60
PTFE	90
Fluorosilicones	240

17.4 Energy costs

Table 17.2 gives estimates of the energy needed to produce 1 kg of a finished material from its raw materials. This includes the energy needed to mine, refine and shape the materials.

Table 17.2 Energy estimates

Material	Energy (MJ/kg)
Metals	
Aluminium and alloys	230 to 300
Copper and alloys	50 to 120
Magnesium and alloys	360 to 420
Iron	20
carbon steels	50 to 60
stainless steels	100 to 120
cast irons	50 to 250
Titanium and alloys	500 to 550
Polymers	
Nylons	170 to 180
Polyethylene	
low density	80 to 100
high density	100 to 120
Polypropylene	110 to 115
Polystyrene	100 to 140
PVC	70 to 90
Synthetic elastomers	120 to 140
Ceramics	
Glasses	10 to 25
Cement	4 to 8
Concrete	3 to 6
Composites	
Glass fibre reinforced polymers	90 to 120
Wood	2 to 4
Reinforced concrete	8 to 20

18 Materials index

Elements

Table 18.1 gives the symbols, atomic mass numbers, atomic numbers and density at 20°C of the elements.

Table 18.1 The elements

Element	Symbol	Atomic mass number	Atomic number	Density (10^3 kg m^{-3})
Actinium	Ac	227	89	10.1
Aluminium	Al	26.98	13	2.7
Americum	Am	243	95	11.8
Antimony	Sb	121.75	51	6.7
Argon	A	39.95	18	1.8(g)
Arsenic	As	74.92	33	5.7
Astatine	At	210	85	
Barium	Ba	137.34	56	3.6
Berkelium	Bk	247	97	
Beryllium	Be	9.01	4	1.9
Bismuth	Bi	208.98	83	9.8
Boron	B	10.81	5	2.5
Bromine	Br	79.91	35	3.1
Cadmium	Cd	112.40	48	8.7
Caesium/Cesium	Cs	132.91	55	1.9
Calcium	Ca	40.08	20	1.5
Californium	Cf	251	98	
Carbon	C	12.01	6	2.3 (graphite)
Cerium	Ce	140.12	58	6.8
Chlorine	Cl	35.45	17	3.2(g)
Chromium	Cr	52.00	24	7.2
Cobalt	Co	58.93	27	8.9
Copper	Cu	63.54	29	8.9
Curium	Cm	247	96	
Dysprosium	Dy	162.50	66	8.5
Einsteinium	Es	254	99	
Erbium	Er	167.26	68	9.1
Europium	Eu	152.0	63	5.2
Fermium	Fm	253	100	
Fluorine	F	19.00	9	1.7(g)
Francium	Fr	223	87	
Gadolinium	Gd	157.25	64	7.9
Gallium	Ga	69.72	31	5.9
Germanium	Ge	72.59	32	5.3
Gold	Au	196.97	79	19.3
Hafnium	Hf	178.49	72	13.3
Helium	He	4.00	2	0.18(g)
Holmium	Ho	164.93	67	8.8
Hydrogen	H	1.008	1	0.09(g)
Indium	In	114.82	49	7.3
Iodine	I	126.90	53	4.9
Iridium	Ir	192.2	77	22.4
Iron	Fe	55.85	26	7.9
Krypton	Kr	83.8	36	3.7(g)
Lanthanum	La	138.91	57	6.2
Lawrencium	Lr	257	103	
Lead	Pb	207.19	82	11.3
Lithium	Li	6.94	3	0.53
Lutetium	Lu	174.97	71	9.8

Magnesium	Mg	24.31	12	1.7
Manganese	Mn	54.94	25	7.4
Mendelevium	Md	256	101	
Mercury	Hg	200.59	80	13.55
Molybdenum	Mo	95.94	42	10.2
Neodymium	Nd	144.24	60	7.0
Neon	Ne	20.18	10	0.90(g)
Neptunium	Np	237	93	19.5
Nickel	Ni	58.71	28	8.9
Niobium	Nb	92.91	41	8.4
Nitrogen	N	14.007	7	1.25(g)
Nobelium	No	254	102	
Osmium	Os	190.2	76	22.5
Oxygen	O	16.00	8	1.43(g)
Palladium	Pd	106.4	46	12.0
Phosphorus	P	30.97	15	1.8
Platinum	Pt	195.09	78	21.45
Plutonium	Pu	242	94	19
Polonium	Po	210	84	9.4
Potassium	K	39.10	19	0.86
Praseodymium	Pr	140.91	59	6.5
Promethium	Pm	147	61	
Protactinium	Pa	231	91	
Radium	Ra	226	88	5
Rhenium	Re	186.2	75	21.0
Rhodium	Rh	102.91	45	12.5
Rubidium	Rb	85.47	37	1.5
Ruthenium	Ru	101.1	44	12.4
Samarium	Sm	150.35	62	7.5
Scandium	Sc	44.96	21	3.1
Selenium	Se	78.96	34	4.8
Silicon	Si	28.09	14	2.3
Silver	Ag	107.87	47	10.5
Sodium	Na	22.99	11	0.97
Strontium	Sr	87.62	38	2.6
Sulphur	S	32.06	16	2.1 (yellow)
Tantalum	Ta	180.95	73	16.6
Technetium	Tc	98	43	11.5
Tellurium	Te	127.60	52	6.2
Terbium	Tb	158.92	65	8.3
Thallium	Tl	204.37	81	11.85
Thorium	Th	232.04	90	11.5
Thulium	Tm	168.93	69	9.3
Tin	Sn	118.69	50	7.3
Titanium	Ti	47.90	22	4.5
Tungsten	W	183.85	74	19.3
Uranium	U	238.03	92	19.05
Vanadium	V	50.94	23	6.1
Xenon	Xe	131.30	54	5.9(g)
Ytterbium	Yb	173.04	70	7.0
Yttrium	Y	88.91	39	4.5
Zinc	Zn	65.37	30	7.1
Zirconium	Zr	91.22	40	6.5

Note: the densities are at 20°C and are for solids unless marked (g) for gas.

Engineering metals

The following is an alphabetical listing of metals, each being listed according to the main alloying element, with their key characteristics. It is not a comprehensive list of all metallic elements, just those commonly encountered in engineering.

Aluminium. Used in commercially pure form and alloyed with copper, manganese, silicon, magnesium, tin and zinc. Alloys exist in two groups: casting alloys and wrought alloys. Some alloys can be heat treated. Aluminium and its alloys have a low density, high electrical and thermal conductivity and excellent corrosion resistance. Tensile strength tends to be of the order of 150 to 400 MPa, with the tensile modulus about 70 GPa. There is a high strength to weight ratio. See Chapter 4.

Beryllium. Beryllium metal is very costly and only used in very special circumstances. Beryllium is more often used as an alloying element, in particular with copper, nickel or steels. Beryllium metal has a high tensile strength and tensile modulus, but is very brittle.

Chromium. Chromium is mainly used as an alloying element in stainless steels, heat-resistant alloys and high strength alloy steels. It is generally used in these for the corrosion and oxidation resistance it confers on the alloys.

Cobalt. Cobalt is widely used as an alloy for magnets, typically 5–35% cobalt with 14–30% nickel, and 6–13% aluminium. Cobalt is also used for alloys which have high strength and hardness at room and high temperatures. These are often referred to as Stellites. Cobalt is also used as an alloying element in steels.

Copper. Copper is very widely used in the commercially pure form and alloyed in the form of brasses, bronzes, cupro–nickels, and nickel silvers. Copper and its alloys have good corrosion resistance, high electrical and thermal conductivity, good machinability, can be joined by soldering, brazing and welding, and generally have good properties at low temperatures. The alloys have tensile strengths ranging from about 180 to 300 MPa and a tensile modulus about 20 to 28 GPa. See Chapter 5.

Gold. Gold is very ductile and readily cold worked. It has good electrical and thermal conductivity.

Iron. The term ferrous alloys is used for the alloys of iron. These alloys include carbon steels, cast irons, alloy steels and stainless steels. See Chapter 3.

Lead. Other than its use in lead storage batteries, it finds a use in lead-tin alloys as a metal solder, and in steels to improve the machinability.

Magnesium. Magnesium is used in engineering alloyed mainly with aluminium, zinc, and manganese. The alloys have a very low density and though tensile strengths are only of the order of 250 MPa there is a high strength to weight ratio. The alloys have a low tensile modulus, about 40 GPa. They have good machinability. See Chapter 6.

Molybdenum. Molybdenum has a high density, high electrical and thermal conductivity and low thermal expansivity. At high temperatures it oxidizes. It is used for electrodes and support members in electronic tubes and light bulbs, and heating elements for furnaces. Molybdenum is however more widely used as an alloying element in steels. In tool steels it improves hardness, in stainless steels it improves corrosion resistance, and in general in steels it improves strength, toughness and wear resistance.

Nickel. Nickel is used as the base metal for a number of alloys with excellent corrosion resistance and strength at high temperatures. The alloys are basically nickel–copper and nickel–chromium–iron. The alloys have tensile strengths between about 350 and 1400 MPa, the tensile modulus being about 220 GPa. See Chapter 7.

Niobium. Niobium has a high melting point, good oxidation resistance and low modulus of elasticity. Niobium alloys are used for high temperature items in turbines and missiles. It is used as an alloying element in steels.

Palladium. This metal is highly resistant to corrosion. It is alloyed with gold, silver or copper, to give metals which are used mainly for electrical contacts.

Platinum. The metal has a high resistance to corrosion, is very ductile and malleable, but expensive. It is widely used in jewellery. Alloyed with elements such as iridium and rhodium, the metal is used in instruments for items requiring a high resistance to corrosion.

Silver. Silver has a high thermal and electrical conductivity, and is very soft and ductile.

Tantalum. Tantalum is a high melting point, highly acid-resistant, very ductile metal. Tantalum-tungsten alloys have high melting points, high corrosion resistance and high tensile strength.

Tin. Tin has a low tensile strength, is fairly soft and can be very easily cut. Tin plate is steel plate coated with tin, the tin conferring good corrosion resistance. Solders are essentially tin alloyed with lead and sometimes antimony. Tin alloyed with copper and antimony gives a material widely used for bearing surfaces. Copper-tin alloys are known as bronzes.

Titanium. Titanium as commercially pure or as an alloy has a high strength coupled with a relatively low density. It retains its properties over a wide temperature range and has excellent corrosion resistance. Tensile strengths are typically of the order of 1100 MPa and tensile modulus about 110 GPa. See Chapter 8.

Tungsten. This is a dense metal with the highest melting point of any metal (3410°C). It is used for light bulb and electronic tube filaments, electrical contacts, and as an alloying element in steels. As whiskers it is used in many metal-whisker composites. See Chapter 11.

Vanadium. This metal is mainly used as an alloying element in steels, e.g. high-speed tool steels.

Zinc. Zinc has very good corrosion resistance and hence finds a use as a coating for steel, the product being called galvanised steel. It has a low melting point and hence zinc alloys are used for products such as small toys, cogs, shafts, door handles, etc. produced by die casting. Zinc alloys are generally about 96% zinc with 4% aluminium and small amounts of other elements or 95% zinc with 4% aluminium, 1% copper and small amount of other elements. Such alloys have tensile strength of about 300 MPa, elongations of about 7–10% and hardness of about 90 BH.

Zirconium. An important use of this metal is as an alloying element with magnesium and steels.

Engineering polymers

The following is an alphabetical listing of the main polymers used in engineering, together with brief notes of their main characteristics. For more details see Chapter 9.

Acrylonitrile butadiene styrene (ABS). ABS is a thermoplastic polymer giving a range of opaque materials with good impact resistance, ductility and moderate tensile and compressive strength. It has a reasonable tensile modulus and hence stiffness, with good chemical resistance.

Acetals. Acetals, i.e. polyacetals, are thermoplastics with properties and applications similar to those of nylons. A high tensile strength is retained with time in a wide range of environments. They have a high tensile modulus and hence stiffness, high impact resistance and a low coefficient of friction. Ultraviolet radiation causes surface damage.

Acrylics. Acrylics are transparent thermoplastics, trade names for such materials including Perspex and Plexiglass. They have high tensile strength and tensile modulus, hence stiffness, good impact resistance and chemical resistance, but a large thermal expansivity.

Butadiene acrylonitrile. This is an elastomer, generally referred to as nitrile or Buna-N rubber (NBR). It has excellent resistance to fuels and oils.

Butadiene styrene. This is an elastomer and is very widely used as a replacement for natural rubber because of its cheapness. It has good wear and weather resistance, good tensile strength, but poor resilience, poor fatigue strength and low resistance to fuels and oils.

Butyl. Butyl, i.e. isobutene-isoprene copolymer, is an elastomer. It is extremely impermeable to gases.

Cellulosics. This term encompasses cellulose acetate, cellulose acetate butyrate, cellulose acetate propionate, cellulose nitrate and ethyl cellulose. All are thermoplastics. Cellulose acetate is a transparent material. Additives are required to improve toughness and heat resistance. Cellulose acetate butyrate is similar to cellulose acetate but less temperature sensitive and with a greater impact strength. Cellulose nitrate colours and becomes brittle on exposure to sunlight. It also burns rapidly. Ethyl cellulose it tough and has low flammability.

Chlorosulphonated polyethylene. This is an elastomer, trade name Hypalon. It has excellent resistance to ozone with good chemical resistance, fatigue and impact properties.

Epoxies. Epoxy resins are, when cured, thermosets. They are frequently used with glass fibres to form composites. Such composites have high strength, of the order of 200–420 MPa, and stiffness, about 21–25 GPa.

Ethylene propylene. This is an elastomer. The copolymer form, EPM, and the terpolymer form, EPDM, have very high resistance to oxygen, ozone and heat.

Ethylene vinyl acetate. This is an elastomer which has good flexibility, impact strength and electrical insulation properties.

Fluorocarbons. These are polymers consisting of fluorine attached to carbon chains. See polytetrafluoroethylene.

Fluorosilicones. See silicone rubbers.

Melamine formaldehyde. This resin, a thermoset, is widely used for impregnating paper to form decorative panels, and as a laminate for table and kitchen unit surfaces. It is also used with fillers for moulding knobs, handles, etc. It has good chemical and water resistance, good colourability and good mechanical strength.

Natural rubber. This is an elastomer. It is inferior to synthetic rubbers in oil and solvent resistance and oxidation resistance. It is attacked by ozone.

Nylons. The term nylon is used for a range of thermoplastic materials having the chemical name of polyamides. A numbering system is used to distinguish between the various forms, the most common engineering ones being nylon 6, nylon 66 and nylon 11. Nylons are translucent materials with high tensile strength and of medium stiffness. Additives such as glass fibres are used to increase strength. Nylons have low coefficients of friction, which can be further reduced by suitable additives. For this reason they are widely used for gears and rollers. All nylons absorb water.

Phenol formaldehyde. This is a thermoset and is mainly used as a reinforced moulding powder. It is low cost, and has good heat resistance, dimensional stability and water resistance.

Polyacetal. See acetals.

Polyamides. See nylons.

Polycarbonates. Polycarbonates are transparent thermoplastics with high impact strength, high tensile strength, high dimensional stability and good chemical resistance. They are moderately stiff. They have good heat resistance and can be used at temperatures up to 120°C.

Polychloroprene. This is an elastomer, usually called neoprene. It has good resistance to oils and good weathering resistance.

Polyesters. Two forms are possible, thermoplastics and thermosets. Thermoplastic polyesters have good dimensional stability, excellent electrical resistivity and are tough. They discolour when subject to ultraviolet radiation. Thermoset polyesters are generally used with glass fibres to form composite materials.

Polyethylene. Polyethylene, or polythene, is a thermoplastic material. There are two main types: low density (LDPE) which has a branched polymer chain structure and high density (HDPE) with linear chains. Materials composed of blends of the two forms are available. LDPE has a fairly low tensile strength and tensile modulus, with HDPE being stronger and stiffer. Both forms have good impermeability to gases and very low absorption rates for water.

Polyethylene terephthalate (PET). This is a thermoplastic polyester. It has good strength and stiffness, is transparent and has good impermeability to gases. It is widely used as bottles for fizzy drinks.

Polypropylene. Polypropylene is a thermoplastic material with a low density, reasonable tensile strength and stiffness. Its properties are similar to those of polyethylene. Additives are used to modify the properties.

Polypropylene oxide. This is an elastomer with excellent impact and tear strengths, good resilience and good mechanical properties.

Polystyrene. Polystyrene is a transparent thermoplastic. It has moderate tensile strength, reasonable stiffness, but is fairly brittle and exposure to sunlight results in yellowing. It is attacked by many solvents. Toughened grades, produced by blending with rubber, have better impact properties.

Polysulphide. This is an elastomer with excellent resistance to oils and solvents, and low permeability to gases. It can however be attacked by micro-organisms.

Polysulphone. This is a strong, comparatively stiff, thermoplastic which can be used to a comparatively high temperature. It has good dimensional stability and low creep.

Polytetrafluoroethylene (PTFE). PTFE is a tough and flexible thermoplastic which can be used over a very wide temperature range. Because other materials will not bond with it, the material is used as a coating to items where non-stick facilities are required.

Polyvinyls. Polyvinyls are thermoplastics and include polyvinyl acetate, polyvinyl butyral, polyvinyl chloride (PVC), chlorinated polyvinyl chloride and vinyl copolymers. Polyvinyl acetate (PVA) is widely used in adhesives and paints. Polyvinyl butyral (PVB) is mainly used as a coating material or adhesive. PVC has high strength and stiffness, being a rigid material. It is frequently combined with plasticizers to give a lower strength, less rigid, material. Chlorinated PVC is hard and rigid with excellent chemical and heat resistance. Vinyl copolymers can give a range of properties according to the constituents and their ratio. A common copolymer is vinyl chloride with vinyl acetate in the ratio 85 to 15. This is a rigid material. A more flexible form has the ratio 95 to 5.

Silicone rubbers. Silicone rubbers or, as they are frequently called, fluorosilicone rubbers, have good resistance to oils, fuels and solvents at high and low temperatures. They do, however, have poor abrasion resistance.

Styrene-butadiene-styrene. This is called a thermoplastic rubber. Its properties are controlled by the ratio of styrene to butadiene. The properties are comparable to those of natural rubber.

Urea formaldehyde. This is a thermosetting material and has similar applications to melamine formaldehyde. Surface hardness is very good. The resin is also used as an adhesive.

Engineering ceramics

The term ceramics covers a wide range of materials and here only a few of the more commonly used engineering ceramics are considered. See Chapter 10 for more details.

Alumina. Alumina, ie. aluminium oxide, is a ceramic which finds a wide variety of uses. It has excellent electrical insulation properties and resistance to hostile environments. Combined with silica it is used as refractory bricks.

Boron. Boron fibres are used as reinforcement in composites with materials such as nickel. See Chapter 11.

Boron nitride. This ceramic is used as an electric insulator.

Carbides. A major use of ceramics is, when bonded with a metal binder to form a composite material, as cemented tips for tools. These are generally referred to as bonded carbides, the ceramics used being generally carbides of chromium, tantalum, titanium and tungsten. See bonded carbides in Chapter 10.

Chromium carbide. See carbides.

Chromium oxide. This ceramic is used as a wear resistant coating.

Glasses. The basic ingredient of most glasses is silica, a ceramic. Glasses tend to have low ductility, a tensile strength which is markedly affected by microscopic defects and surface scratches, low thermal expansivity and conductivity (and hence poor resistance to thermal shock), good resistance to chemicals and good electrical insulation properties. Glass fibres are frequently used in composites with polymeric materials. See Chapter 11.

Kaolinite. This ceramic is a mixture of aluminium and silicon oxides, being a clay.

Magnesia. Magnesia, ie. magnesium oxide, is a ceramic and is used to produce a brick called a dolomite refractory.

Pyrex. This is a heat resistant glass, being made with silica, limestone and boric oxide. See glasses.

Silica. Silica forms the basis of a large variety of ceramics. It is, for example, combined with alumina to form refractory bricks and with magnesium ions to form asbestos. It is the basis of most glasses.

Silicon nitride. This ceramic is used as the fibre in reinforced materials, such as epoxies. See Chapter 11.

Soda glass. This is the common window glass, being made from a mixture of silica, limestone and soda ash. See glasses.

Tantalum carbide. See carbides.

Titanium carbide. See carbides.

Tungsten carbide. See carbides.

19 Selection of processes

19.1 *Processes and choice of material*

This chapter is a consideration of the characteristics of the various types of processes which can be used to form products. In making a decision about the manufacturing process to be used there are a number of questions that have to be answered:

1 *What is the material?*
   The type of material to be used affects the choice of processing method. Thus, for example, if casting is to be used and the material has a high melting point then the process must be either sand casting or investment casting.

2 *What is the shape?*
   The shape of the product is generally a vital factor in determining which type of process can be used. Thus, for example, a product in the form of a tube could be produced by centrifugal casting, drawing or extrusion but not generally by other methods.

3 *What type of detail is involved?*
   Is the product to have holes, threads, inserts, hollow sections, fine detail, etc.? Thus, for example, forging could not be used if there was a requirement for hollow sections.

4 *What dimensional accuracy and tolerances are required?*
   High accuracy would rule out sand casting, though investment casting might well be suitable.

5 *Are any finishing processes to be used?*
   Is the process used to give the product its final finished state or will there have to be an extra finishing process. Thus, for example, planing will not produce as smooth a surface as grinding.

6 *What quantities are involved?*
   Is the product a one-off, a small batch, a large batch or continuous production? While some processes are economic for small quantities, others do not become economic until large quantities are involved.

19.2 *Surface finish*

Roughness is defined as the irregularities in the surface texture which are inherent in the production process but excluding waviness and errors of form. Roughness takes the form of a series of peaks and valleys which may vary in both height and spacing and is a characteristic of the process used. Waviness may arise from such factors as machine or work deflections, vibrations, heat treatment or warping strains. Roughness and waviness may be superimposed on departures of the surface from the true geometrical form.

Waviness

Roughness

Waveiness combined
with roughness

Figure 19.1 Surface finish

One measure of roughness is the *arithmetical mean deviation*, denoted by the symbol R_a. This is the arithmetical average of the variation of the profile above and below a reference line throughout the prescribed sampling length. The reference line may be the centre line, this being a line chosen so that the sums of the areas contained between it and those parts of the surface profile which lie on either side of it are equal.

$$R_a = \frac{\text{sum of areas A + sum of areas B}}{\text{sample length}} \times 1000$$

where the sample length is in millimetres and the areas in square millimetres.

Table 19.1 Relation of R_a values to surface texture

Surface texture	Roughness R_a μm
Very rough	50
Rough	25
Semi-rough	12.5
Medium	6.3
Semi-fine	3.2
Fine	1.6
Coarse-ground	0.8
Medium-ground	0.4
Fine-ground	0.2
Super-fine	0.1

Table 19.2 Typical roughness values for different processes

Process	Roughness R_a μm
Sand casting	25 to 12.5
Hot rolling	25 to 12.5
Sawing	25 to 3.2
Planing, shaping	25 to 0.8
Forging	12.5 to 3.2
Milling	6.3 to 0.8
Drilling	6.3 to 0.8
Chemical milling	6.3 to 0.8
Boring, turning	6.3 to 0.4
Investment casting	3.2 to 1.6
Extruding	3.2 to 0.8
Cold rolling	3.2 to 0.8
Drawing	3.2 to 0.8
Die casting	1.6 to 0.8
Grinding	1.6 to 0.1
Honing	0.8 to 0.1
Electrolytic grinding	0.6 to 0.2
Polishing	0.4 to 0.1

19.3 *Characteristics of metal forming processes*

Casting of metals

Casting can be used for components from masses of about 10^{-3} kg to 10^4 kg, with wall thicknesses from about 0.5 mm to 1 m. Castings need to have rounded corners, no abrupt changes in section and gradual sloping surfaces. Casting is likely to be the optimum method in the circumstances listed below but not for components that are simple enough to be extruded or deep drawn.

1   *The part has a large internal cavity*
    There would be a considerable amount of metal to be removed if machining was used; casting removes this need.

2   *The part has a complex internal cavity*
    Machining might be impossible; by casing, however, very complex internal cavities can be produced.

3   *The part is made of a material which is difficult to machine*
    The hardness of a material may make machining very difficult, e.g. white cast iron, but this presents no problem with casting.

4   *The metal used is expensive and so there is to be little waste*
    Machining is likely to produce more waste than occurs with casting.

5   *The directional properties of a material are to be minimised*
    Metals subject to a manipulative process often have properties which differ in different directions.

6   *The component has a complex shape*
    Casting may be more economical than assembling a number of individual parts.

An important consideration in deciding whether to use casting and which casting process to use is the tooling cost for making the moulds. When many identical castings are required, a method employing a mould which may be used many times will enable the mould cost to be spread over many items and may make the process economic. Where just a one-off product is required, the mould used must be as cheap as possible since the entire cost will be defrayed against the single product. Table 18.3 shows the main characteristics of the various casting processes.

The following factors largely determine the type of casting process used:

1   *Large heavy casting*
    Sand casting can be used for very large castings.

2   *Complex design*
    Sand casting is the most flexible method and can be used for very complex castings.

3   *Thin walls*
    Investment casting or pressure die casting can cope with walls as thin as 1 mm. Sand casting cannot cope with such thin walls.

4   *Good reproduction of detail*
    Pressure die casting or investment casting gives good reproduction of detail, sand casting being the worst.

Table 19.3 Casting processes

Process	Usual materials	Section thickness mm	Size kg
Sand casting	Most	> 4	0.1–200 000
Gravity die casting	Non-ferrous	3 to 50	0.1–200
Investment casting	All	1 to 75	0.005–700
Centrifugal casting	Most	3 to 100	25 mm–1.8 m diameter
Pressure die casting: high pressure	Non-ferrous	1 to 8	0.0001–5
Pressure die casting: low pressure	Non-ferrous	2 to 10	0.1–200

Table 19.3 Continued

Process	Size kg	Roughness R_a μm	Production rate, items per hour
Sand casting	0.1–200 000	25 to 12.5	1–60
Gravity die casting	0.1–200	3.2 to 1.6	5–100
Investment casting	0.005–700	3.2 to 1.6	Up to 1000
Centrifugal casting	25 mm–1.8 m diameter	25 to 12.5	Up to 50
Pressure die casting: high pressure	0.0001–5	1.6 to 0.8	Up to 200
Pressure die casting: low pressure	0.1–200	1.6 to 0.8	Up to 200

Good surface finish
Pressure die casting or investment casting gives the best finish, sand casting being the worst.
High melting point alloys
Sand casting or investment casting can be used.
Tooling cost
This is highest with pressure die casting. Sand casting is cheapest. However, with large number production, the tooling costs for metal moulds can be defrayed over a large number of castings, whereas the cost of the mould for sand casting is the same no matter how many castings are made since a new mould is required for each casting.

Table 19.4 Typical applications of casting

Process	Applications
Sand casting	Engine blocks, machine tool bases, pump housings
Gravity die casting	Cylinder heads, pistons, gear and die blanks
Investment casting	Machine tool parts, turbine blades, engine components
Centrifugal casting	Pipes, brake drums, pulley wheels, gun barrels
Pressure die casting	Transmission cases, domestic appliances, pump components

Manipulation of metals

Manipulative methods involve the shaping of a material by means of plastic deformation methods. Such methods include forging, extruding, rolling, and drawing. Depending on the method, components can be produced from as small as about 10^{-5} kg to 100 kg, with wall thicknesses from about 0.1 mm to 1 m. Table 19.5 shows typical process characteristics. Table 19.6 shows typical uses of products formed from such processes.

Table 19.5 Manipulative processes

Process	Usual materials	Section thickness mm	Minimum size
Closed-die forging	Steels, Al, Cu, Mg alloys	3 upwards	10 cm²
Roll forming	Any ductile material	0.2 to 6	
Drawing	Any ductile material	0.1 to 25	3 mm dia.
Impact extrusion	Any ductile material	0.1 to 20	6 mm dia.
Hot extrusion	Most ductile materials	1 to 100	8 mm dia.
Cold extrusion	Most ductile materials	0.1 to 100	8 mm dia

Note: ductile materials are commonly aluminium copper an magnesium alloys and to a lesser extent carbon steels an titanium alloys.

Table 19.5 Continued

Process	Maximum size	Roughness R_a μm	Production rate, item/hour
Closed-die forging	7000 cm²	3.2 to 12.5	Up to 300
Roll forming		0.8 to 3.2	
Drawing	6 m dia.	0.8 to 3.2	Up to 3000
Impact extrusion	0.15 m dia.	0.8 to 3.2	Up to 2000
Hot extrusion	500 mm dia.	0.8 to 3.2	Up to 720 m
Cold extrusion	4 m long	0.8 to 3.2	Up to 720 m

Compared with casting, wrought products tend to have a greater degree of uniformity and reliability of mechanical properties. The manipulative processes do, however, tend to give a directionality of properties which is not the case with casting. Manipulative processes are likely to be the optimum method for product production when:

1. *The part is to be formed from sheet metal*
   Depending on the form required, shearing, bending or drawing may be appropriate if the components are not too large.

2. *Long lengths of constant cross-section are required*
   Extrusion or rolling would be the optimum methods in that long lengths of quite complex cross-section can be produced without any need for machining.

3. *The part has no internal cavities*
   Forging can be used when there are no internal cavities, particularly if better toughness and impact strength is required than are obtainable with casting. Also, directional properties can be imparted to the material to improve its performance in service.

4. *Seamless cup-shaped objects or cans are required*
   Deep drawing or impact extrusion would be optimum methods.

5. *The component is to be made from material in wire or bar form*
   Bending or upsetting can be used.

Table 19.6 Typical applications of forming processes

Process	Application
Closed-die forging	Connecting rods, crank shafts
Roll forming	Bar, sheet, window frame sections
Drawing	Cans, kitchen appliances, electrical fittings
Impact extrusion	Fasteners, tool sockets
Extrusion	Window frames, decorative trim

Powder processes

Powder processes enable large numbers of small components to be made at high rates of production, for small items up to 1800 per hour, and with little, if any, finishing machining required. It enables components to be made with all materials and in particular with those which otherwise cannot easily be otherwise easily processed, e.g. the high melting point metals of molybdenum, tantalum and tungsten, and where there is a need for some specific degree of porosity, e.g. porous bearings to be oil filled. The mechanical compaction of powders only, however, permits two-dimensional shapes to be produced, unlike casting and forging. In addition the shapes are restricted to those that are capable of being ejected from the die. Thus, for example, reverse tapers, undercuts and holes at right angles to the pressing direction have to be avoided. On a weight-for-weight basis, powdered metals are more expensive than metals for use in manipulative or casting processes; however, this higher cost may be offset by the absence of scrap, the elimination of finishing machining and the high rates of production. The maximum size for products is about 4.5 kg.

Machining metals

When selecting a cutting process, the following factors are relevant in determining the optimum process or processes:

1. Operations should be devised so that the minimum amount of material is removed. This reduces material costs, energy costs involved in the machining and costs due to tool wear.
2. The time spent on the operation should be kept to a minimum to keep labour costs low.
3. The skills required affect the labour being costs.
4. The properties of the material being machined should be considered; in particular the hardness. In general, the harder a material the longer it will take to cut. The hardness also, however, affects the choice of tool material that can be used and, in the case of very hard materials, the process that can be used. Thus, for instance, grinding is a process that can be used with very hard materials because the tool material, the abrasive particles, can be very hard. Where a considerable amount of machining occurs, the use of free-machining grades of materials should be considered as a means of minimising cutting times.
5. The process, or processes, chosen should take into account the quantity of products required and the required rate of production.
6. The geometric form of the product should be considered in choosing the most appropriate process or processes.
7. The required surface finish and dimensional accuracy also affect the choice of process or processes.

Machining operations vary quite significantly in cost, particularly if the operation is considered in terms of the cost necessary for a particular operation to achieve particular tolerances. For example, to achieve a tolerance of 0.10 mm, the rank order of the processes is:

Shaping	Most expensive
Planing	
Horizontal boring	
Milling	
Turret (capstan)	Least expensive

The cost of all processes increases as the required tolerance is increased. At high tolerances, grinding is one of the cheapest processes. As Table 19.7 indicates, the different machining operations also produce different surface finishes.

Table 19.7 Surface finishes

Machining process	$R_a \mu$m	
Planing and shaping	25 to 0.8	Least smooth
Drilling	8 to 1.6	
Milling	6.3 to 0.8	
Turning	6.3 to 0.4	
Grinding	1.6 to 0.1	Most smooth

The choice of process will also depend on the geometric form required for the product. Table 19.8 indicates the processes that can be used for different geometric forms.

Table 19.8 Machining processes for particular geometric forms

Type of surface	Suitable process
Plane surface	Shaping, planing, face milling, surface grinding
Externally cylindrical surface	Turning, grinding
Internally cylindrical surface	Drilling, boring, grinding
Flat and contoured surfaces and slots	Milling, grinding

Machining, in general, is a relatively expensive process when compared with many other methods of forming materials. The machining process is, however, a very flexible process which allows the generation of a wide variety of forms. A significant part of the total machining cost of a product is due to setting-up times when there is a change from one machining step to another. By reducing the number of machining steps and hence the number of setting-up times, a significant saving becomes possible. Thus the careful sequencing of machining operations and the choice of machine to be used is important.

Joining processes with metals

Fabrication involves the joining of materials and enables very large structures to be assembled, much larger than can be obtained by other methods such as casting or forging. The main joining processes are essentially adhesive bonding, soldering and brazing, welding and fastening systems. The factors that determine the joining process are the materials involved, the shape of the components being joined, whether the joint is to be permanent or temporary, limitations imposed by the environment and cost. Welded, brazed and adhesive joints, and some fastening joints, e.g. riveted, are generally meant to be permanent joints, while soldered joints and bolted joints are readily taken apart and rejoined.

19.4 *Characteristics of polymer-forming processes*

Injection moulding and extrusion are the most widely used processes. Injection moulding is generally used for the mass production of small items, often with intricate shapes. Extrusion is used for products which are required in continuous lengths or which are fabricated from materials of constant cross-section. The following are some of the factors involved in choosing a process:

1   *Rate of production*
    Typical cycle times: injection moulding and blow 10–60 s, compression moulding (cycle time 20–600 s), rotational moulding 70– 1200 s, thermoforming (cycle time 10–60 s).

Table 19.9 Joining processes with metals

Process	Materials	Min.-max. sheet thickness mm	Production rate m/min
Tungsten inert-gas welding	Most non-ferrous, carbon & low alloy steels	0.1–3	0.2 manual 1.5 auto.
Metal inert-gas welding	Most non-ferrous, carbon, low alloy & stainless steels	0.5–80	0.5 manual
Manual metal arc welding	Carbon, low alloy & stainless steels, nickel alloys	1.5–200	0.2
Submerged arc welding	Carbon, low alloy & stainless steels	5–20	5
Resistance welding	Generally low carbon steels, not cast iron, or high carbon steel	0.3–6	
Gas welding	Generally ferrous alloys	0.5–30	0.1
Brazing	Most metals	0.1–50	High
Soldering	Most metals	0.1–6	
Adhesives	Most metals	0.05–50	High
Mechanical fasteners	Most metals	0.25–200	Can be high

2 *Capital investment required*
   Injection moulding requires the highest capital investment, with extrusion and blow moulding requiring less capital. Rotational moulding, compression moulding, transfer moulding, thermoforming and casting require the least capital investment.

3 *Most economic production run*
   Injection moulding, extrusion and blow moulding are economic only with large production runs. Thermoforming, rotational moulding and machining are used with small production runs. Table 19.10 indicates the minimum output that is likely to be required to make processes economic.

4 *Surface finish*
   Injection moulding, blow moulding, rotational moulding, thermoforming, transfer and compression moulding, and casting all give very good surface finishes. Extrusion gives only a fairly good surface finish.

5 *Metals inserts during the process*
   These are possible with injection moulding, rotational moulding, transfer moulding and casting.

6 *Dimensional accuracy*
   Injection moulding and transfer moulding are very good, compression moulding and casting good, extrusion poor.

Table 19.10 Minimum output

Process	Economic output number
Machining	From 1 to 100 items
Rotational moulding	From 100 to 1000 items
Sheet forming	From 100 to 1000 items
Extrusion	Length 300 to 3000 m
Blow moulding	From 1000 to 10 000 items
Injection moulding	From 10 000 to 100 000 items

7   *Item size*
    Injection moulding and machining are the best for very small items. Section thicknesses of the order of 1 mm can be obtained with injection moulding, forming and extrusion.

8   *Enclosed hollow shapes*
    Blow moulding and rotational moulding can be used.

9   *Intricate, complex shapes*
    Injection moulding, blow moulding, transfer moulding and casting can be used.

10 *Threads*
    Threads can be produced with injection moulding, blow moulding, casting and machining.

11 *Large formed sheets*
    Thermoforming can be used.

Table 19.11 shows the processing methods that are used for some commonly used thermoplastics and Table 19.12 for thermosets.

Table 19.11 Processing methods for thermoplastics

Polymer	Extrusion	Injection moulding	Extrusion blow moulding	Rotational moulding
ABS	*	*		*
Acrylic	*	*		
Cellulosics	*	*		
Polyacetal	*	*	*	
Polyamide	*	*		*
Polycarbonate	*	*	*	
Polyester	*	*		
Polyethylene HD	*	*	*	*
Polyethylene LD	*	*	*	*
Polyethylene terephthalate	*	*	*	
Polypropylene	*	*	*	
Polystyrene	*	*	*	*
Polysulphone	*	*		
PTFE	*			
PVC	*	*	*	*

Table 19.11 continued overpage

Table 19.11 Continued

Polymer	Thermo-forming	Casting	Bending & joining	As film
ABS	*		*	
Acrylic	*	*	*	
Cellulosics	*			*
Polyacetal				
Polyamide		*		*
Polycarbonate	*		*	
Polyester				
Polyethylene HD			*	*
Polyethylene LD			*	*
Polyethylene terephthalate				*
Polypropylene	*		*	*
Polystyrene	*		*	*
Polysulphone	*			
PTFE				
PVC	*		*	*

Table 19.12 Processing methods for thermosets

Polymer	Compression moulding	Transfer moulding	Casting
Epoxy			*
Melamine formaldehyde	*	*	
Phenol formaldehyde	*	*	*
Polyester	*	*	*
Urea formaldehyde	*	*	

Table 19.12 Continued

Polymer	Laminate	Foam	Film
Epoxy	*	*	
Melamine formaldehyde	*		
Phenol formaldehyde	*	*	
Polyester	*		*
Urea formaldehyde	*		

Assembly processes that can be used with plastics are welding, adhesive bonding, riveting, press and snap-fits, and thread systems.

19.5 Costs for processes

The manufacturing cost, for any process, can be considered to be made up of two elements – fixed costs and variable costs. Figure 19.2 shows the typical forms of the graphs of costs against quantity for processes where in (a) there is a low fixed cost but high variable cost per item, e.g. sand casting, and for where in (b) there is a high fixed cost but low variable cost per item, e.g. die casting. The total cost is the sum of the fixed cost plus the cost per item and Figure 19.3 shows the two total costs lines for the processes giving the graphs in Figure 19.1 (a) and (b). Below the quantity N, the process giving (a) is cheaper than that giving (b); above that (b) is cheaper than (a).

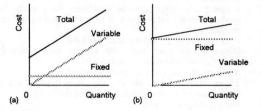

Figure 19.2 Costs

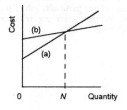

Figure 19.3 Costs

The cost elements are made up as follows:

1 *Fixed costs*

   Capital costs for installations, e.g. the cost of a machine or even a foundry, are usually defrayed over an expected lifetime of the installation. However, in any one year there will be depreciation of the asset and this is the capital cost that is defrayed against the output of the product in that year. Thus if the capital expenditure needed to purchase a machine and install it was, say, £50 000, then in one year a depreciation of 10% might be used and thus £5000 defrayed as capital cost, i.e. an element of the fixed cost, against the quantity of product produced in that year. Another element of fixed costs is the cost of dies or tools needed specifically for the product concerned. Other factors we could include in the fixed costs are plant maintenance and tool or die overhaul.

2 *Variable costs*

   The variable costs are the material costs, the labour costs, the power costs, and any finishing costs required.

The questions to be asked when costing a product are:

1   *Is the installation to be used solely for the product concerned?*
    The purpose of this question is to determine whether the entire capital cost has to be written off against the product or whether it can be spread over a number of products.

2   *Is the tooling to be used solely for the product concerned?*
    If specific tooling has to be developed for a particular product then the entire cost will have to be put against the product.

3   *What are the direct labour costs per item?*
    Direct labour costs are those of the labour directly concerned with the production process.

4   *What are the other labour costs involved?*
    These, termed indirect labour costs, include such costs as those incurred in supervision, inspection, etc.

5   *What is the power cost?*

6   *Are there any finishing processes required and, if so, what are their costs?*

7   *What are the materials costs?*

8   *Are there any overhead costs to be included?*
    Overhead costs are those costs which a company cannot specifically allocate to any particular job or product but are part of the overall company costs, e.g. telephone costs, rent for factory, management salaries.

Appendix: Units

The International System (SI) of units has seven base units:

Length	metre	m
Mass	kilogram	kg
Time	second	s
Electric current	ampere	A
Temperature	kelvin	K
Luminous intensity	candela	cd
Amount of substance	mole	mol

In addition there are two supplementary units, the radian and the steradian.

The SI units for other physical quantities are formed from the base units via the equation defining the quantity concerned. Density is defined by the equation density = mass/volume and thus has the unit of mass divided by the unit of volume, i.e. kg/m^3. Some of the derived units are given special names. For example, the unit of force is defined by the equation force = mass × acceleration and is thus kg m/s or kg m s^{-1} and given the name newton (N). The unit of stress is given by the equation stress = force/area and has the derived unit of N/m^2; this is given the name pascal (Pa) and so 1 Pa = 1 N/m^2. Certain quantities are defined as the ratio of two quantities with the same units. For example, strain is defined as change in length/length and is thus expressed as a pure number with no units.

Standard prefixes are used for multiples and submultiples of units, the SI preferred ones being multiples of 10^3. Table A.1 gives details. For example, 1000 N can be written as 1 kN, 1 000 000 Pa as 1 MPa, 1 000 000 000 Pa as 1 GPa, 0.001 m as 1 mm, and 0.000 001 A as 1 μA. Note that often the unit N/mm^2 is used for stress, 1 N/mm^2 is 1 MPa.

Other units which the reader may come across are fps (foot-pound-second) units which still are often used in the USA. On that system the unit of length is the foot (ft), with 1 ft = 0.3048 m. The unit of mass is the pound (lb), with 1 lb = 0.4536 kg. The unit of time is the second, the same as the SI system. With this system the derived unit of force, which is given a special name, is the poundal (pdl), with 1 pdl = 0.1383 N. However, a more common unit of force is the pound force (lbf). This is the gravitational force acting on a mass of 1 lb and consequently, since the standard value of the acceleration due to gravity is 32.174 ft/s^2, then 1 lbf = 32.174 pdl = 4.448 N. The similar unit the kilogram force (kgf) is sometimes used. This is the gravitational force acting on a mass of 1 kg and consequently, since the standard value of the acceleration due to gravity is 9.806 65 m/s^2, then 1 kgf = 9.806 65 N. A unit often used for stress in the USA is the psi, or pound force per square inch. 1 psi = 6.895 × 10^3 Pa. Table A.2 gives conversion factors from the fps to SI system and vice versa.

Table A.1 Standard prefixes for SI units

Multiplication factor	Prefix		
1 000 000 000 000 000 000 000 000 =	yotta	10^{24}	Y
1 000 000 000 000 000 000 000 =	zetta	10^{21}	Z
1 000 000 000 000 000 000 =	exa	10^{18}	E
1 000 000 000 000 000 =	peta	10^{15}	P
1 000 000 000 000 =	tera	10^{12}	T
1 000 000 000 =	giga	10^{9}	G
1 000 000 =	mega	10^{6}	M
1 000 =	kilo	10^{3}	k
100 =	hecto	10^{2}	h
10 =	deca	10	da
0.1 =	deci	10^{-1}	d
0.01 =	centi	10^{-2}	c
0.001 =	milli	10^{-3}	m
0.000 001 =	micro	10^{-6}	μ
0.000 000 001 =	nano	10^{-9}	n
0.000 000 000 001 =	pico	10^{-12}	p
0.000 000 000 000 001 =	femto	10^{-15}	f
0.000 000 000 000 000 001 =	atto	10^{-18}	a
0.000 000 000 000 000 000 001 =	zepto	10^{-21}	z
0.000 000 000 000 000 000 000 001 =	yocto	10^{-24}	y

Table A.2 Conversion factors

Length	1 m = 3.28 ft , 1 cm = 0.394 in
	1 ft = 30.48 cm, 1 in = 2.54 cm
Area	1 cm^2 = 0.155 in^2
	1 in^2 = 6.452 cm^2
Volume	1 m^3 = 35.315 ft^2, 1 cm^3 = 0.0610 in^3
	1 ft^3 = 0.0283 m^3, 1 in^3 = 16.39 cm^3
Mass	1 kg = 2.205 lb
	1 lb = 0.454 kg
Density	1 kg m^{-3} = 0.0624 lb ft^{-3}, 1 Mkg m^{-3} = 0.0361 lb in^{-3}
	1 lb ft^{-3} = 16.02 kg m^{-3}, 1 lb in^{-3} = 27.68 Mg m^{-3}
Force	1 N = 7.233 pdl, 1 N = 0.2248 lbf
	1 pdl = 0.1383 N, 1 lb f = 4.448 M
	1 kg f = 9.807 N
Stress	1 Pa = 0.0209 lbf ft^{-2} = 145.0 lbf in^{-3}
	1 lbf ft^{-2} = 47.880 Pa, 1 lbf in^{-2} = 6.895 kPa
Energy	1 J = 0.7376 ft lbf = 23.730 ft pdl = 0.2388 cal
	1 cal = 4.187 J, 1 ft pdl = 0.0421 J, 1 ft lbf = 1.356 J
	1 btu = 1.055 kJ
Power	1 kW = 1.341 hp
	1 hp = 550 ft lbf s^{-1} = 0.7457 kW
Temp.	1 K = 1°C = 1.8 °F, t°C = 273.15 + t K
	1 °F = 5/9 K, $(t$°F − 32)/9 = t°C/5
Thermal	1 W m^{-1} K$^-$ = 2.39 % 10^{-3} cal cm^{-1} K^{-1}
conduct.	1 cal cm^{-1} K^{-1} = 4.18 % 10^{-2} W m^{-1} K^{-1}

Hardness

Table A.3 shows the approximate relationship between the main hardness scales when used with steels, Table A.4 when used with non-ferrous alloys.

Table A.3 Hardness scales for steels

Brinell 300 kg 10 mm ball	Rockwell		Vickers diamond pyramid
	A	C	
615	81.3	60.1	700
585	80.0	57.8	650
550	78.6	55.2	600
512	77.0	52.3	550
471	75.3	49.1	500
425	73.3	45.3	450
379	70.8	40.8	400
331	68.1	35.5	350
284	65.2	29.8	300
238	61.6	22.2	250
190			200
143			150
95			100

Table A.4 Hardness scales for non-ferrous alloys

Brinell 300 kg 10 mm ball	Rockwell B	Vickers diamond pyramid
190	93.8	200
181	91.6	190
171	89.2	180
162	86.5	170
152	83.4	160
143	80.0	150
133	76.1	140
124	71.5	130
114	66.3	120
105	60.0	110
95	52.5	100

Index

This index does not list all the terms used, for these the reader is referred to Chapter 1 where there is an alphabetical listing. The index also does not list all the various forms of alloys and polymers, only the main categories being listed. For full details the reader is referred to Chapter 18 where each commonly used metal and polymer, in alphabetical order, is briefly discussed and to the chapters specifically concerned with polymers or the metal.

231